S0-CAG-640

No Vacancy

No Vacancy

Global Responses to the Human Population Explosion

Introduction by Michael Tobias, Ph.D.
Edited by Michael Tobias, Bob Gillespie,
Elizabeth Hughes and Jane Gray Morrison

Hope Publishing House
Pasadena, California

For information address:

Hope Publishing House
P.O. Box 60008
Pasadena, CA 91116 - U.S.A.
Tel: (626) 792-6123 / Fax: (626) 792-2121
E-mail: hopepub@sbcglobal.net
Web sites: http://www.hope-pub.com
www.novacancythemovie.org
Printed on acid-free paper

Cover design — Michael McClary/The Workshop

Library of Congress Cataloging-in-Publication Data

No vacancy : global responses to the human population explosion / introduction by Michael Tobias ; edited by Michael Tobias ... [et al].
p. cm.
Includes bibliographical references.
ISBN 13: 978-1-932717-08-2 (trade pbk. : alk. paper)
ISBN 10: 1-932717-08-0 (trade pbk. : alk. paper)
1. Population policy. 2. Fertility, Human. 3. Overpopulation. I. Tobias, Michael.
HB883.5.N62 2005
363.9--dc22

2005031526

Contents

Introduction

The problem

In the time of Christ, humans numbered between 200 and 300 million. By the mid-19th century, we had reached 1.5 billion people. By 1963 that population had doubled. Today the human population exceeds 6.4 billion.

Scientists, demographers, family planners and ecologists track these trends. Thomas Malthus' theory is confirmed by the evident patterns that suggest an exponential increase in human numbers and the inability of our food production – certainly with meat-based diets – to keep apace. Because of the increasing technological success reflected by enormous agricultural output, particularly in the wake of the Green Revolution of the 1950s and 1960s, our understanding of the perils of rapid population growth has been somewhat muted by many enthusiasts who would argue that technology and human ingenuity will always come to the rescue. And if not, then Nature will, in her own way, in her own time, stabilize our numbers.

Many people find such cold calculus inhumane, lacking the resilience of sophisticated social science, of fundamental human rights, of forward-thinking public policy and underlying empathy. However, consider the numerous issues related to population.

There are huge differences of opinion among professionals and the public alike. The 21st century places challenges before humankind like no other period in our history. Our very biological triumph as a species, traditionally defined by reproductive success, is poised to backfire, as billions are born into a world where parents, communities and governments cannot adequately protect their livelihoods or basic needs. There may be an excitement about the rising middle class in India. However there is even greater alarm at the ever-expanding gulf between rich and poor Indians. Half the world's population lives on less than $4 a day and is currently having family sizes that, in some regions, will double or triple the existing population size.

The number of environmental refugees in the world is increasing, as people flee natural disasters, disappearing watersheds, shortage of arable land and logged-out forests. Displaced in a new physical setting they forget the techniques for putting bread on their tables. In China, the most populous nation in human history, water shortages are becoming critical, as is the surplus of unemployed or underpaid workers. Continuing poverty across much of South Asia and Africa places an untenable burden upon wildlife. The AIDS pandemic has impacted the traditional transmission of agricultural knowledge because parents die of AIDS before they can teach their children to farm.

Throughout the world, more and more species are going extinct, while whole ecosystems are sapped of their integrity and robustness. An unprecedented burden is being placed upon the biosphere by an ever-growing number of human consumers in rich and poor countries alike. The economic imbalances are instructive, and need strong corrections, but the end-losers are not merely the nearly 750 million hungry children or the women who still suffer under the weight of male dominance, but all of Nature.

According to most demographers – some whose voices are heard in this book – our species is likely to *stabilize* (to stop growing in number) somewhere between 9 and 13 billion. But those are projections based upon all couples in the world eventually deciding to have replacement size families. Whether at 9 or 13 billion, as one begins to try and calculate the consumptive impact such numbers impose upon an already fast-diminishing natural resource base, the consequences of such demographic growth take on stark reality. Every reader of this book, every couple planning its family, every medical professional, scientist, educator and policy maker the world over has the challenge to create a sustainable world for future generations of children. Each of us needs to be accountable for our impact on the environment.

The Solutions

For all of the caveats and dire warnings described above, there is–on a far brighter note–a true fertility revolution occurring today in various parts of the world. Many of these trends are described here by the very professionals, healthcare workers and scientists who have devoted their lives to finding solutions to rapid population growth. Lamentable problems, particularly for women and children, have traditionally resulted from inadequate government policies, insufficient delivery of services and poor levels of information available to countless women and their partners.

When family-planning programs began in the early 1960s in the Far East and India, over 90% of the fieldworkers were women, as were their supervisors and administrative staff. Now there are literally millions of door-to-door family-planning fieldworkers who tirelessly promote health and family planning at peoples' doorsteps. This book is a result of interviews with hundreds of professionals who have had a leadership role in providing maternal and child health services, family planning, development and environmental protection. These initiatives are all ultimately linked.

In Iran, we learned how the Islamic clerics were able to develop primary health programs and contraceptive services that resulted in family sizes declining from six to close to replacement size level (approximately 2.1 children per couple) in just 15 years.

In India we interviewed feminist leaders and early pioneers and heard the dynamic journalist Menakshee Shedde describe the challenges and opportunities of achieving successful health- and family-planning programs. We discovered details about the active participation of women in small, empowering groups known as *Mandalas* and how women have increasingly been able to control their political, financial and educational futures.

In Indonesia we met with the founding leaders and current directors of similar services and discovered what have come to be the common threads of all successful programs. Indonesians have dramatically lowered their birth- and death-rates in what is the fourth largest nation and the largest Muslim population in the world.

In Mexico the political commitment to a national program for delivering contraceptive services was begun in the mid-1970s and coördinated by a multi-ministerial population council with the active participation of the ministry of health, the extensive social security hospitals and health centers. The private sector has also had a significant role in this revolution by dint of its widespread involvement through the channels of private physicians, nongovernmental organizations, women's groups and the social marketing of condoms and oral pills.

As this book reveals, throughout the world, population experts provide a wealth of experience and insight. At UCLA, Professor Ana Maria Goldani describes the complexity of fertility factors and interactions from different regions in Brazil and explains why couples have small families as a result of women's empowerment plus the increasing cost of the third child in marginally subsistent slums as well as in the tropical rainforest. In 1950 when Brazil had 52 million people, a population growing at an annual rate of 3% and a total fertility rate of 6.2, a Brazilian could expect to live to be 44. Today, life expectancy is 69 and the country's total fertility rate is 2.2 with its growth rate down to 1.3%.

Brazilian social scientist Dr. Carmen Barroso, past director of the MacArthur Foundation's Population Program and current executive director of the Western Hemisphere of International Planned Parenthood, offers a deep level of understanding about such remarkable transformations, describing how programs have been oriented towards preventing adolescent pregnancy, improving the role and status of women and providing a social climate that enables women to participate in the labor market.

In Africa, a continent of huge challenges and discrepancies, we chose Ghana and Nigeria for reasons of instructive contrast. Both nations have important programs in place, but by their differences and similarities they illustrate the issues facing most African nations. In Ghana, we encounter the dramatically successful Navrongo Health Research Program which impacts the medical and family practices throughout the country. Dr. James Phillips of the Population Council, a principal architect of the program, organized meetings in New York with project leaders and directors of the Ghanaian health service, and also facilitated capturing the program visually in Navrongo, a northern province of Ghana.

Nigeria is a most complex, diverse and challenging country. The current population of 134 million people is projected to reach over 300 million in 50 years. The current fertility rate is 5.8 and only 18% of the women currently practice contraception, while less than 1% of the gross national product is spent on education and only 2% goes to healthcare. What has worked in Ghana might work in Nigeria, but it will take time.

In Western Europe, the population dilemmas are not about too many mouths to feed, or women and children to protect, but are about decreased fertility. Dozens of Europeans were interviewed on the streets of Holland, France and Italy to glean some sense of the public's thinking on fertility issues. The concerns of an aging population are very real throughout Europe (as well as in Japan and some other countries) and the social welfare nets to deal with the elderly will inevitably come under huge pressure with the workforce shrinking. This is a mismatch for which few current policies exist. What is also spectacularly true about Europe is the transformation that has occurred for women. They have a level of independence and access to healthcare and to the work place unthinkable a few generations ago.

In the United States, all the individuals interviewed are highly distinguished professionals. Some of them direct policies and programs for the World Bank, the United Nations Fund for Population Activities, the Office of Population and Reproductive Health at USAID and the United Nations Population Division. Others work for foundations such as Hewlett, Packard, Weeden and Alfred P. Sloan; some are attached to family planning, population and environmental organizations which include the Western Hemisphere Region of the IPPF (International Planned Parenthood Federation), the World Watch Institute, the Heinz Center for Science, Economics and the Environment, the Earth Policy Institute, Conservation International, the Center for Entrepreneurship in International Health and Development, the Planned Parenthood Federation of America, the Population Council, Population Action International and Marie Stopes International. Other distinguished scholars were interviewed on various campuses, including Malcolm Potts at the University of California Berkeley, Paul Ehrlich at Stanford University and Allan Rosenfield at Columbia University.

The problems described by the experts in these fields in the U.S. are especially complex given America's great wealth and how sharply its citizens disagree over as abortion and immigration. The bipartisan United States Commission on Immigration Reform (USCIR) submitted a comprehensive report on immigration reform policies to Congress. The USCIR was chaired by Shirley Hufstedler, whose voice rings loud and clear in this book, as does that of Michael Teitelbaum, a member of the commission who has had a long and distinguished career both as a foundation director and immigration specialist.

The ecological perspective of numerous and brilliant scientists in the U.S. lends a comprehensive overview to the challenges facing this country and its responsibilities to the rest of the world. The biosphere clearly cannot continue to be plundered by an ever-increasing number of human consumers, in the U.S. or elsewhere. This is the ecological and practical reality that has provided the conceptual umbrella for this book, *No Vacancy.*

A Collective Wisdom

While it would not be possible in this introduction to refer to each and every interviewee, the editors do thank all those who graciously endured the reality of a film documentary by which means these interviews were conducted. Every speaker in this book sat under the hot glare of film lights and the often distracting medium itself, which sometimes calls for retakes, new angles, tedious delays, technical equipment problems and hassles. But, the end result, we believe, makes for a much more accessible, candid, conversational, unedited conveyance for the general reader than might otherwise be imagined.

From the outset we must stress that these are *not* academic essays, heavily footnoted. Rather they are discussions held in front of camera crews at work. The editors have tried to maintain the distinctive voice of each speaker, as opposed to forcing a unity of tone and style. Many contributions in the book have been truncated (in some cases to a tenth of their length in the transcripts). The collective wisdom resulting from this documentary approach, we think, provides readers with a unique insight into the feelings as well as the profound information base these extraordinary individuals bring from their expertise and lifelong experience.

That collective wisdom seems to confirm there have been substantive changes for the better within the population arena worldwide. Some 45% of all nations have now reached fertility replacement. There is much work yet to be done. The United Nations Population Conferences in Belgrade (1964), Bucharest (1974), Mexico City (1984) and Cairo (1994) revealed an ever-increasing commitment by governments, donors, foundations and NGOs to help couples prevent unwanted pregnancies and to integrate family planning within the context of maternal and child health. The common thread to all of these endeavors is the empowerment of women.

Over the last two generations, professionals in the field have witnessed a rapid decline in infant mortality, an increase in the age at marriage, and greater birth spacing–all crucial ingredients of smaller family sizes. Moreover, the value of small families has been reinforced by women's coöperatives, their access to credit (note the extraordinary success, for example, of the Grameen Bank), of literacy programs and education for youth, especially young girls.

This book explores the broad framework of population and development policies with emphasis on female education, literacy programs, technical assistance and the many trials and tribulations that continue to characterize the struggle of the healthcare community to provide safe motherhood and to ensure that every child is a wanted child. All these areas combined collectively impact the global environment in both obvious and more subtle ways. Clearly many aspects of those women's health and family-planning programs that work are virtually universal.

These include utilizing all qualified health professionals to provide a cafeteria of choices with methods stressed according to the needs of couples. Services are provided by traditional birth attendants, public and private physicians, and full-time family-planning fieldworkers. Social marketing programs deliver condoms and oral pills in public outlets (i.e., roadside kiosks) alongside tea, soap, matches and cigarettes. All successful programs use mass media and entertainment. Other key ingredients include decentralization that empowers local government ministries, including legislators and the judiciary.

Surveys of couples in their reproductive age and the analysis of clinical records by dedicated fieldworkers provide guidance in expanding and improving services. Sex education in schools, as seen in Mexico, is also a crucial component that is increasingly successful in country after country. Improving the status of women, perhaps the most important element of all, comes through in these interviews. Professor J.J. van Dam says, "We have focused for so long on safety and family size that we often forget sex has often to do with love between two people as well as the love for their children."

The editors hope that readers will take the time to explore all of the many issues raised throughout *No Vacancy* in greater detail. An attached appendix provides some (but by no means all) of the important organizations and websites doing fantastic work in these areas throughout the world.

Acknowledgements

This book, the film and its distribution were funded by the Fred H. Bixby Foundation. The Bixby Foundation has as its mandate funding innovative family planning/women's reproductive health and population organizations. Additionally, the Dancing Star Foundation, and Population Communication contributed substantially to this book, and the accompanying feature film documentary by the same title.

The Dancing Star Foundation, founded in 1993, is a nonprofit public benefit corporation devoted to animal welfare, global biodiversity conservation and environmental education. The foundation's primary emphasis is the maintenance of various sanctuaries which provide a refuge for wild habitat and species, as well as domestic animals. In addition, the foundation is involved in population and ecological research and analysis, book publications such books as *No Vacancy*, as well as various other lectures, colloquia and environmental filmmaking.

Population Communication was founded in 1977 to determine the policies and programs needed to achieve population stabilization. It encourages world leaders to support the *Statement on Population Stabilization* which has been signed by 75 heads of government. In addition Population Communication develops and tests cradle-to-grave child survival, adolescent health, birth spacing and small family policies and programs in seven countries. By contracting with a manufacturer in Taiwan, Population Communication donates maternal- and child-health medical equipment to 41 countries.

–*Michael Tobias,* Ph.D.

1

Iran

Introduction

In 1987 under the dynamic leadership of Dr. Seyed Alireza Marandi, Iran's minister of health, a dialogue was initiated to explore the potential of reinvigorating a family planning and maternal and child health program. He was supported by longtime family-planning pioneers, such as Dr. Amir H. Mehryar who had been director of the Population Center at Shiraz University but now directs the Institute for Research on Development and Planning. Numerous articles by brilliant scholars such as Dr. Mohammad Jalal Abbasi-Shavazi, assistant professor in the department of demography and social sciences at the University of Tehran, substantiated the need to address the effect of rapid population growth on development as well as the devastating impact on women's health.

What resulted was a massive health mobilization and reconstruction program. Within a mere 15 years, birthrates had dropped to replacement-size families. Infant mortality was reduced by two-thirds and maternal mortality dropped to a fourth of its prior level. Under the dynamic leadership of Dr. Bahram Delavar, director general of the family health department in the Ministry of Health, over 15,000 "Health Houses" provided primary healthcare, with an extensive network of over 4,500 health centers. Over 100,000 health professionals were mobilized in both the rural and urban areas and 5,000 rural midwives were trained. Two decades ago there were seven teaching hospitals in universities but now there are 53 hospitals where universities graduate doctors.

Close to 90% of the entire population throughout the country is visited by 23,000 male and female family-planning health workers who offer them a total array of contraceptive services, referring both men and women for surgical sterilization as well as injectables and implants. A total of 220,000 men have obtained vasectomies. Currently 72% of all couples practice family planning.

Dr. Hossein Malek-Afzali

Family Planning in Iran

Iran's Family Planning Program is administered by the ministry of health with services offered to three target groups. Programs for adolescents covering various family-planning issues are taught in schools in coördination with the ministry of education. Training courses instruct soldiers about issues pertaining to reproductive health and family planning. In order to graduate, university students are required to study and pass classes in family planning and population control.

Family-planning instruction is also given to those planning on marriage because by law no couples will be issued marriage certificates until they have completed a required class in family planning. Initially, training programs dealt exclusively with family planning but now they have been expended to include counseling on sexual, social and emotional relations in marriage.

The third target group are those who are married and must decide on family planning. This group receives the majority of all family-planning services. Married women, aged 15 to 49, are eligible for these services.

Behvarzes are trained healthcare workers, both male and female, who provide primary healthcare services at the village level.

Each *Behvarz* works through a Health House which is situated in a village and easily accessible to a population of about 1,500. *Behvarzes* also make house visits and offer a wide variety of services: antenatal and postnatal care, school health, environmental health in the workplace, child vaccinations, growth monitoring and nutrition education. Family planning and counseling is offered to women between 10 and 49 years of age. The *Behvarz* can refer patients needing greater care to a Rural Health Center (RHC) run by a doctor and health technicians. The *Behvarz* thus frees the medical staff from remedial health concerns and keeps the RHC available for serious issues requiring higher training and skills.

In 1983, the Ministry of Health began offering short training courses to rural midwives to improve the delivery of services in areas with no maternal hospitals. Each trainee was known locally and trusted by the indigenous population. The midwives were also given equipment to use under the supervision of the *Behvarz* who was required to be present during delivery of births. At the end of 2001, the number of trained midwives had reached 8,300.

The urban equivalent of the Health House and Rural Health Center are the Health Posts and Urban Health Centers. Here the average population served by each location is 12,000. Beyond these there are District Health Centers (DHC) where the next level of medical care is offered by a staff of specialists and experts. The DHC reports to the provincial university of medical sciences and health Services, which in turn reports to the ministry of health and medical education. Family-planning policy is issued at the ministry level, but plans and recommendations come from the field up through the system.

Dr. Hossein Malek-Afzali is a deputy health minister and Dr. Bahram Delavar is general director of the Family Health and Population Department, both at the Ministry of Health responsible for policy-making and planning for the reproductive health and family-planning programs in Iran [Ministry of Health, Treatment and Medical Education, Hafez Ave. and Jomhoury St., Tehran, Iran, tel: (98 21) 670-675, 693-4226, fax: (98 21) 671-872].

Seyed Alireza Marandi

Health Aspects of Many Pregnancies

In the name of God the compassionate and merciful. Following the Islamic revolution, the family-planning program was automatically halted. It was a general belief among everybody, including the people in charge of the ministry of health, that family planning was contrary to Islamic ideology. The general impression was that we needed more children, more young people, so religious leaders concluded that when Prophet Mohammad, peace be on him, started his work, he promoted having more Muslims, and therefore we should continue this idea forever.

I wrote a letter to the late imam pointing out the health aspects of women with repeated pregnancies, advocating the need for birth spacing and forewarning there would be more and more fatalities and complications with unplanned pregnancies. To everybody's surprise, including my own, Ayatollah Khomeini, peace be upon him, responded to this letter in public proclaiming the population issue to be a vital one that should be discussed in public, in the media, as well as in academia, to see what should be done about it. With this encouragement given by Ayatollah Khomeini, dialogue began in the media and the universities. Those in favor of family planning were able to air their views and soon they were implemented because there were few opponents to these plans.

Eventually the populace understood family planning as something religious leaders and our ideology accepted and in certain situations encouraged. We fielded mobile teams to travel throughout the country and established other facilities where we could not only educate the populace, but also offer them tubal ligations or vasectomies. No one was coerced, yet all over Iran family planning became available – free services, according to the law, for any who wanted to take advantage of them. Our population growth rate declined from 3.9% in 1986 to 1.4% in 1996.

Pediatrician Seyed Alireza Marandi, professor at Shahid Beheshti University of Medical Sciences and Health Services, formerly was Iran's minister of health. He won the 1999 United Nations Population Award and the Dr. A.T. Shousha Foundation Prize from WHO in 2000 [Shahid Beheshti University of Medical Sciences and Health Services, Eveen Ave., Teheran 19395 4719, Iran, tel: (98 21) 841-9393, 840-2049].

Dr. Amir Mehryar

The Revolution

The first year after the revolution, there were efforts by the ministry of health to keep the family-planning program alive. Health officials not only went to the late Imam Khomeini to get a *fatwah* from him in favor of contraception, they also approached several other ayatollahs. After the 1986 census, we suddenly became aware that our rate of population growth was quite high, around 3.9%. At first the government was delighted our population was increasing so rapidly. At the time we were at war and the only advantage we had against Iraq, which was getting support from other countries like the United States, was our population. In a few years, after the war ceased leaving our economy in bad shape, our officials realized a high rate of population growth was unsustainable and so Iran began a series of activities to call attention to the pressures and consequences of over-population.

The first step was a population seminar organized jointly by a few universities and the ministry of planned budget in Mashad. The discussion, at rather academic levels, explored the effect of population on health, education, government spending, food supplies, basic needs, and so on. This prepared the ground for government action, and by the end of this session the ministry of health announced Iran was going to adopt a family-planning program.

People, feeling the crunch of economic stress, found it difficult to support a large family and at the same time make any progress in their family's status and welfare, so they embraced family planning warmly. A survey in 1988, before the program was fully implemented, showed that nearly half of all Iranians were already using family planning. Numerous contraceptives were made available–pills, condoms, IUDs and, of course, there was always the traditional method of withdrawal.

Interestingly enough, Iran, a government headed by fundamentalist Muslims, was ready to initiate family planning and offer sterilization to men and women despite the fact such methods are not approved by many sects in other Muslim countries. In this respect Iran is singular. The Shiite clerics have been more liberal from the beginning and this has helped the Iranian family-planning program. In Iran, female sterilization is the second most popular method for those who want to take control of their lives and undergo sterilization rather than wait for their husbands to do it; however, male sterilization is gaining in popularity among men in urban areas.

Dr. Amir H. Mehryar is the director of the Institute for Research on Development and Planning. His papers, "The Evolution of Family Planning and Recent Fertility Decline in the Islamic Republic of Iran" and "Iranian Miracle: How to Raise Contraceptive Prevalence Rate to above 70% and Cut TFR by Two-thirds in less than a Decade," describe the social and demographic transformation that has taken place in Iran [Institute for Research on Planning and Development (IRDP), PO Box 19395/4647, Tehran, Iran, tel: (98 21) 229-0062, 229-0063].

Dr. Professor Mohammad Jalal Abbasi-Shavazi

Iran's Fertility Decline

Iran has experienced a phenomenal decline in population growth recently–the speed of which is equal to any seen in the world. Whether in rural or urban areas, all provinces of Iran have experienced this same decline. Recently a survey titled "The Iran Fertility Transition" was conducted with the collaboration of the demography program of the Australian National University, the

University of Iran and the Ministry of Health. We covered four provinces, including Yaz and Gilan plus two rather remote provinces, Sistan and Baluchistan. We wanted to see the trend and level of fertility in these provinces as well as to explain the causes for these trends. Fertility has declined phenomenally from 6.5 in 1986 to around 2.0 in the year 2000 – a remarkable decline for an Islamic country. In fact, I think the Iran experience in population control is unique for all the countries in the Middle East.

In Iran after the revolution, the status of women improved considerably – mainly by an increase in the level of education. In remote provinces such as Baluchistan, we asked women if they preferred their daughters to continue their education or to get married at an earlier age after they finished high school. More than 70% of married women preferred their daughters to continue their education.

Dr. Mohammad Jalal Abbasi-Shavazi, professor and chair of the department of sociology and demography at the University of Tehran, is the lead Iranian expert to the Population Division at the United Nations Secretariat and is one of the most active researchers in the field of demography in Iran. He has done postdoctoral studies in Australia, is an associate of the demography program at the Australian National University and is one of the founding members and the secretary general of the Population Association of Iran [Dept. of Demography, Faculty of Social Sciences, University of Tehran, 16 Azar Street, Enghelab Ave., Tehran, Iran, tel: (98 21) 802-4936, fax: (98 21) 640-9348. www.ut.ac.ir].

The Family Planning Association of the Islamic Republic of Iran

The Family Planning Association (FPA) of the Islamic Republic of Iran, a member of IPPF, complements the services offered by the government by providing information and mobilizing support for reproductive health and family planning among religious and political leaders and the public.

The Youth Committee (YC) project was established in 1996. Members vote for four male and female members below the age of 25 every two years. The YC disseminates peer family-planning education through workshops, conferences and seminars; topics include AIDS prevention and sexual and reproductive rights. The structure of the YC allows for workshop attendees to be trained as instructors to replace older members as they age out of the project.

[Family Planning Association of the Islamic Republic of Iran, PO Box 19395-3518, Tehran 19119, Iran, tel: (98 21) 222-3944, 222-1479, fax: (98 21) 225-7746, e-mail: info@fpairi.org, www.fpairi.org].

Zahra Majdfar

Iran's Health Network

By 2001, some 8,300 rural midwives were working in Iran's health network. They were equipped to properly assist deliveries. Some 32,000 Behvarz Health Workers serve at 16,000 Health Houses throughout the country. The Rural Health Centers number nearly 3,000, covering 95% of the rural population. In the cities, since there is a sizable active private sector complementing the National Health Network, over 95% of the urban population is covered. Health Posts refer needy patients to urban Health Centers, which, in turn, refer them to hospitals within the National Health Network.

A family-planning administrator at Shahid Beheshti University, Ms. Zahra Majdfar has a master's degree in midwifery and is in charge of the population control and family-planning program at SBU [Shahid Beheshti University of Medical Sciences and Health Services, Eveen Ave., Teheran 19395 4719, Iran, tel: (98 21) 841-9393, 840-2049].

Mullah Mohammad Hossein Shamsian

Islam and Family Planning

In the name of God, the Beneficent and the Merciful, family planning might seem undesirable to some religious people and from some religious points of view, but from an Islamic point of view, given that the Prophet Mohammad himself had said that the fewer the children, the better the family life, I feel that this precept is an intuitive truth.

Comparing countries with low populations to those with higher populations makes this clear. Having a smaller population allows for better care of the population. The same holds true at the family level. A family with fewer members can be cared for better, despite cultural aggression and ethnic degradation. The fewer the children, the better.

The matter of withdrawal is a key issue in the Shiite view of family planning. There is a difference between Shiite and Sunni viewpoints regarding this matter which stems from the fact that Sunni canons are unchangeable, set at some time in the past, while Shiite *Sharia* canons can evolve through reason. The Koran and the collective consent of Muslims are considered in all current decisions. The matter of family planning from a Shiite point of view can include the use of condoms, vasectomies and the pill.

I always felt that health matters and family planning were important, and that having numerous children could be detrimental to the health of the family, particularly from an ethical and moral perspective. I decided to get a vasectomy, thinking it was appropriate to model that method. I am a preacher and as a person who has had a vasectomy, I feel that I can be more effective advocating family planning.

2

India

Introduction

Before visiting India, we contacted the secretary of family welfare in the ministry of health and family planning, Shri J.V. Prasada Rao, and his very enthusiastic joint secretary, J. Sujatha Rao, to explore the status of current family welfare policies and ask why the states of Kerala, Andhra Pradesh and Tamil Nadu have achieved replacement level fertility. We then contacted leaders such as Gopi Gopalakrishnan of Janani in Bihar and state officials in Rajasthan and Uttar Pradesh to explore how we might capture visually the contrast between southern India and the Bimaru states in the Ganges River basin.

Especially helpful were Francois Farah, the UNFPA representative; Saroj Pachauri, the regional director of the population council stationed in New Delhi; Kalpana Apte, director of training for the Family Planning Association of India; Sudha Tewari, director of Parivar Seva Sanstha; Dr. Manmohan Sharma of the Indian Association of Parliamentarians on Population and Development; Dr. Srinivasan, director of the Population Foundation of India; Dr. Spade of Engender; Chandrasekar of Ipas; and Victor Barbiero and Randy Kolstad, population advisors at USAID. S. Ramasundaram was instrumental in decentralizing the target-based program in Tamil Nadu and pioneering AIDS prevention programs. In all, 123 professionals in the government, donor, foundation and NGO communities were contacted.

We wanted to discover what the common elements of successful family planning programs were and what the challenges are in a country that adds 1.5 million new people every month. We asked about current population policies and goals and how decentralization at the block, panchayat (village council) and district level work. What are the leadership and bureaucratic obstacles? What is the progress of training auxiliary nurse midwives (ANMs), private physicians, village midwives and mobilizing social marketing systems? What TV soap operas, motion picture films and traditional entertainment forms are being used to promote family planning and women's health? We viewed "Hari Bhari," directed by Shyam Benegal, a feature film chronicling the stories of five women who are confronted by the dramatic choices related to everything from preventing violence in the family to forced (arranged) marriages.

Feminist leaders described for us the women's literacy and education programs aimed at increasing the age at marriage and exploring the local and policy implications of women's rights. Visits to women's *mandalas,* or women's coöperatives let us observe how they deliver credit, promote literacy, generate income projects and teach women improved agricultural practices. We also visited various private and public clinics, birth and registration offices, manufacturing industries and youth clubs and then talked with family welfare promoters, auxiliary nurse midwives and the shopkeepers who distribute condoms and oral pills.

We were looking for the exciting programs, such as the surgical contraceptive services pioneered at the Pearl Clinic in Mumbai. We talked with the dramatic motion picture director Sai Paranjpye, president of the Children's Film Society of India, who explained how the family welfare is conveyed to youth in schools and brothels.

The theme in all our interviews was to learn the story from the cradle to the grave about what is helping to lower infant mortality, prevent adolescent pregnancy, increase the age at marriage, improve birth spacing, and provide contraceptive and reproductive health services so couples are encouraged and want to have replacement level fertility. We wanted to discover the policy, leadership, education, financial, religious, political, economic, cultural, geographic and ecological considerations involved.

In Haryana and Rajasthan, Sudha Tewari, director of Parivar Seva Sanstha, organized filming of male and female health assistants, medical officers, multipurpose workers, village health guides, and TBA (traditional birth attendants), as well as activities in postpartum hospitals, private clinics, social marketing, factories and adolescent health programs.

Saroj Pachauri

The Population Council's research agenda in India is designed to respond to evolving national needs. The program portfolio is conceptualized under three major themes: population policy, reproductive health and gender, family dynamics and development. Research has been used to further comprehensive health and gender issues. In the arena of adolescent health, the council addresses the challenge faced by first-time parents whose needs are quite different than those of older couples. Recent research on HIV/AIDS focused on commercial sex workers, STD transmission rates, preventative practices and the problems of stigma and discrimination that often keep patients from seeking treatment and counseling.

Prior to Saroj Pachauri's becoming regional director of the Population Council's South and East Asia Office, she served at the Ford Foundation in India. She acknowledges the inspiration for her work has been her mother, one of the first female doctors in India during the 1920s who began the maternal/child health and family-planning program in New Delhi for the urban poor and ran this program for 20 years. Her mother's devoted service to the oppressed Indian classes was the seed that blossomed into Ms. Pachauri's own career.

The Paradigm Shift of Panchayats

For decades in India we had a target-driven, male-dominated, bureaucratic program which focused on numbers of sterilizations and that was what drove the program. Now we want a comprehensive approach to reproductive health, one that addresses the multiple reproductive health needs of people. This represents a paradigm shift. The government of India, fortunately, has pursued it in whatever ways it can.

The issues of informed choice, of reproductive choice, and addressing needs that were felt by the people themselves were the principles on which this whole agenda was built. And this agenda has been linked to the larger process in development that was under way. The panchayats and the democratic decentralization processes involved listening to women and developing programs based on needs of people rather than on needs perceived by the bureaucracy. This requires allowing people at the village level and the poorest of the poor–and women are often the poorest of the poor–to have a say in their own needs. This is happening today with women in various panchayats where their voices are finally being heard–at least in some panchayats. After all, addressing reproductive health needs fit into the panchayat movement which aims to address the development needs of the people.

Studies done around the country showed women suffered a disproportionate load of reproductive tract morbidity and reproductive tract illnesses. When these morbidity and mortality rates continued high, finally the leaders began to face up to needed changes. There was the delicate issue of male partnership which is hard to address in a patriarchal society. Talking about the problems women suffer from is not a priority in such instances.

Gender inequality still looms as an issue that must be addressed. You can demonstrate the problems women have, you can work with women for empowerment, but in my opinion the only way we are going to make headway in these issues is to get both women and men involved. Men in particular *are* the responsible sexual partners. Decisions on reproductive and sexual matters in this country, as in much of the world, are made by men, and so their involvement in this arena is absolutely critical.

The Colossus That Is India

Gender equality is not going to happen overnight because in India gender disparities are huge. The disparities are not the same in different areas of the country as a whole–as between a state like Kerala and states like Andhra Pradesh or Rajasthan. So we cannot discuss India as a unified whole. So, too, gender disparities, as with almost all development issues, vary enormously depending on the area you are in. In Rajasthan, women's empowerment is a goal that is barely on the charts. Women's literacy levels are low, infant mortality rates are high. Male wishes prevail, so the age of marriage is low and fertility rates are high. In Kerala it's a whole different world with infant mortality rates probably lower than those in New York City. In states with low literacy levels, a poor infrastructure and inadequate healthcare services, the unmet need for contraception is high.

Population issues cannot be addressed in isolation. If we were to talk only about family planning, we could make available contraceptives and let it go at that. Since they want contraception and want to limit family size (surveys indicate that generally people don't want large families, even the poor in rural tribal areas do not want large families) they need to choose to limit family size. But such an idea is too radical. To be effective, we need to give them access to services in ways that are engineered so that they can use contraceptives properly.

The Population Council and the Indian Council of Medical Research have been involved in many of the clinical trials on various contraceptive methods, including injectables and Norplant. Presently we don't have a contraceptive preference, bit that is our goal. With the new national population policy giving reproductive choices to both women and men, we need to expand the basket of options so people have choices, especially young people who have not completed their family size and therefore are uninterested in sterilization–currently the main option in the program.

At the same time, these methods require plenty of other services to go with them–including counseling and careful medical examinations of those who cannot use contraceptives because of contra-indications. Quality-of-care standards must be put in place and follow-up is also needed to monitor for adverse side effects. When this is done–and there are many examples in this country where it has been accomplished, especially in the NGO sector–this program should be very effective. Throughout India we must stress that couples need to delay their first pregnancies and then space their children appropriately. Here sterilization is not the answer, so other choices must be made available.

My Mother

Mother became a doctor in the 1920s–the first woman doctor in this country. She was a role model for me–and for many others–as she served the poor and women all her life. I, too, have always had a passion to work on development and health issues with the poor, helping those who are suppressed and oppressed–because that's what I'm trained for. A career devoted to the reproductive health of women, therefore, came very naturally to me.

Saroj Pachauri is the regional director of the South and East Asia office of the Population Council [Zone 5A, Ground Floor, India Habitat Centre, Lodi Rd., New Delhi 110003, India, tel: (91 11 2) 464-2901/2, 4008/9, 465-2502/3, 465-6119, fax: (91 11 2) 464-2903, e-mail: monica@pcindia.org].

Poonam Muttreja

MacArthur Foundation India

The MacArthur Foundation is involved throughout India. Its fellowship program funds mid-management individuals from NGOs, the academic community and government in programs that expand their skills and help them network with other professionals in their field. Grants to NGOs allow such groups to move beyond providing services to start advocacy projects aimed at legal changes as well as addressing the issues of women's rights. Programs for youth will target sexuality and reproductive health education for both in-school and out-of-school adolescents. Women's reproductive health programs include delaying the first birth, birth spacing, and emergency obstetric care coupled with emergency transportation for underserved areas.

Poonam Muttreja, the country coördinator for the MacArthur Foundation in India, remembers having had three grandmothers because each would die in childbirth. With currently available technology, she sees no reason for high maternal mortality rates or the need for women to use abortion as a contraceptive means. To decrease fertility and to empower women around the world is not a dream; it's a reality that can be achieved in the next decade.

Women in India

In a country where the healthcare system shows very little empathy and respect for women, changes must be made so women can gain greater control of their lives. It is unfair that women in India have so few choices about the contraceptive methods they might prefer. Every contraceptive has a problem. There are risks with pills as well as with permanent methods, but the biggest risk for women in India is undergoing an abortion–yet there are more than ten million abortions performed in India each year. Although no one approves of abortion as a method of contraception, that is what it effectively has become.

Women in India deserve to have a choice and control their contraceptive methods because they need safer provisions. Unfortunately, an "injectable" is a choice denied to the women of India because as yet it cannot be provided in the same controlled manner in which it is made available in developed countries. We need to review how this can happen in our trials. We also must strengthen the primary healthcare system. There are two government facilities every country's citizens encounter: the schools and the primary healthcare system. The primary healthcare system is one place all go – young and old. To strengthen these institutions in terms of not just facilities, but to include greater participation of the community, we must develop a transparency in the system.

Another problem which needs further attention is getting an accurate diagnosis and analysis of the population. There are too many myths and misconceptions. We are poor because we have too many people. Most women want no more than two children, at best three, but for that to happen, they must be given confidence those children will survive, so we must invest in neonatal survival, recognizing that if people feel secure about the health of their children and themselves, fertility will decline, maternal mortality will decline and morbidity will decline. The primary healthcare center is the temple where all of this will happen.

Millions of women in India have five children, four children, but virtually no access to reproductive healthcare. While it may be a pleasure to have children, there is a considerable burden in having a large family in terms of household work, economic pressures and education. Many women would prefer to decrease their fertility, yet we are doing little about this issue. We are asking the wrong questions, ignoring that this is all doable. Bangladesh has shown us. Iran has shown us. Decreasing fertility and helping women around the world is not a dream – it's a reality that can be accomplished in India in the next decade.

We must get on with this business and make it happen. There

is plenty to do and the evidence from across the world assures us that we can be successful in this effort.

Poonam Muttreja is country coördinator for MacArthur Foundation India, [Zone VA, 1st floor, India Habitat Centre, Lodi Rd., New Delhi 110 003, India, tel: (91 11) 2464-4006, fax: (91 11) 2464-4007, e-mail: pmuttreja@macfound.org.in, info@macfound.org.in, www.macfound.org.in].

Mrs. Datta Pai, *widow of Dr. Datta Pai, founder of the Pearl Centre*

The Pearl Centre, founded in 1974, was the first private clinic for family planning in Mumbai. It offers abortion and sterilization in addition to in vitro fertilization. In order to make vasectomies more accessible, Dr. Datta Pai organized traveling vasectomy clinics which were set up in railway stations or in a roaming bus that would stop at select locations to perform the operation. The founder's son and daughter-in-law, Drs. Hrishikesh Pai and Rishma Dhillon Pai, carry on the work of the Pearl Centre.

The Pearl Centre

Situated in the heart of the city, the Pearl Centre is truly a pearl because it is a landmark in services for women. In 1972 India passed a law enabling women to access abortion services more widely and the Pearl Centre made it possible for women of Mumbai to exercise that right. Today, this clinic continues to provide services for the whole family – children can get immunized, women can get an abortion or choose various family-planning options. The staff is gentle, the quality of services is high, the facility is spotless and the patients are not intimidated here.

Even though abortions and sterilizations were the first services the Centre offered, a comprehensive family healthcare approach soon evolved as efforts were made to provide care for the entire family's welfare. Besides medical procedures, counseling, pre-abortion counseling and post-abortion counseling were integrated into the program in order to provide better care for the patient.

At the same time everything remained subsidized, so when a patient comes in, the first one to interview this person is an administrator. The patient is then examined by the medical personnel who get an informed consent. If the issue is a family planning procedure, the woman can then decide, with her husband, which method they would like to use. The clinic stresses, by presenting both the pros and cons of the methods offered, that the patients understand their options in order to make informed choices. Coercion of any sort is never part of the program. Some 98% of all clients are followed up. This assures that we build continuity of service and trust between generations. In fact, we are now so well known in Mumbai, we no longer need to advertise.

Dr. Rishma Pai

Abortion and Family Planning

Abortion still remains a large part of family planning. No one claims it is a preferred method of family planning, but unfortunately we often don't see women until they are pregnant – and come to us for abortions. When we do their abortions, we then follow up with education about contraception, because otherwise they won't come into the clinic.

Changing Sexual Mores

In Mumbai, as throughout India, there has been a huge change in attitudes and standards of morality. Where traditionally premarital sex was never countenanced, today it has become a way of life amongst the younger generation. So younger women are coming in with pregnancies, which in earlier times never happened. Young people are also less embarrassed or inhibited about such pregnancies. Whereas in prior times should a young girl become pregnant, she would always pretend she was married because of the social stigma. Today young women come in announcing they are single and pregnant and need to do something about it. It surprises us that even these educated college girls only appear in our clinic when they need an abortion. It makes no sense that they don't use preventative means to keep from getting pregnant. So we tell them – this has happened, just make sure it doesn't happen again.

The Need for Population Education

In our country, most of the population does not go to high school – the appropriate time to get young people into sex education programs. That's why we need community-based sex education programs where educational seminars can be held on weekends in public places to discuss the many problems facing our society – pollution, the environment and family planning – since they are all interrelated. Unfortunately, population issues do not make the headlines, even though they should – since they are the root of all of our problems, whether pollution, disease, AIDS, tuberculosis or the economy.

Dr. Rishma Dhillon Pai is a consultant gynecologist and obstetrician at Jaslok Hospital, Pearl Centre. Recently she opened the Everywoman Clinic to handle gynecological, sexual, psychological and marital problems, as well as general surgery [Pearl Centre, Senapati Bapat Rd., Dara (W), Mumbai 400028 India, tel: (91 22) 430-5278, 430-5823].

Sai Paranjpye

Sai Paranjpye is an independent filmmaker who has long championed women's rights and environmental awareness. Her latest foray has been into the world of children's cinema, an arena long lacking, she believes, in the extensive Indian world of film. She is the chair of the Children's Film Society, India, and continues to create accessible, intelligent films for children.

Her mother, Madam P., an early pioneer in family planning, is profiled in Ms. Paranjpye's documentary, *Persuasion*. At a time when discussing such matters was considered taboo and bad taste

(1940s), Mrs. P. would recruit family-planning advocates from lower-income groups to promote family planning among their peers. At first she met with resistance, but her grassroots approach gradually broke down the barriers to discussing family planning.

Madam P

When she began her work in the 1940s, family planning was something one did not discuss in polite society. Even her own family members were ashamed and considered telling people about birth control to be disgusting. As a Cambridge graduate in mathematics, she felt there was a crying need to spread the word. It was most unusual for a woman to be educated in those years. Most women didn't even finish schooling, they got married. Instead she had finished her degree and married a White Russian, my father.

When my mother began her mission, she knew that were she to go to a village, the villagers would be suspicious of a city woman telling them what to do–and would not accept her as a credible authority. So to circumvent this skepticism she employed a rather unique modus operandi, gathering together a band of about ten people, both men and women, from the poorest areas. She included a sweeper, a hospital worker, a bus conductor, a load carrier and such common laborers. The one thing they all had in common was they had all undergone sterilization operations.

My mother would enter villages with her little band of brave souls and set up a platform that would allow them talk to the villagers. As a kid, I trailed along and thoroughly enjoyed being spoiled rotten by these gracious villagers who sat listening to the load carrier who got up and said, "Look, you've already got six daughters now. You've got to get yourself operated upon." Most were too uninformed to know what was required, so then someone would describe the operation, how simple it was, and the load carrier would assure them that he immediately returned to work the following day.

My mother carried on her mission all over India. But when such pioneers pass away, we lose that fighting spirit. And then our nation suffered a setback with the Emergency.

My film *Papeha*, roughly translated in English as *The Forest Lovebird*, is what I like to call an adult fairy tale. It is a love triangle between a man, a woman and a forest. The prince is a forest officer while the princess is a social anthropologist studying tribal groups. They meet in a forest and, of course, fall in love. An old tribal man tells this story to a group of some eight tribal children.

Through the story-telling the point is made of the necessity, urgency, and dire need for tribal groups to preserve their habitat and to flourish in their own surroundings. For me it was a wonderful opportunity to work with tribal groups. There are more than 300 different types of tribal groups in India. All the inputs were absolutely true to their way of life, their problems, their needs, their beliefs, and everything that went into it was based around that. When you destroy an ecosystem, you destroy the human beings who survive within that system, so it has been a struggle for these tribes trying to lead their own life, trying to be children of nature for as long as they could.

Another of my films tells of two friends who come to Bombay–one a pessimist who claims Bombay is a city of no return, like a lion's cage; the other, newly married, has left his heart in his village and says since his roots are in the village, he will return there one day. This film examines the tension between the village and the city, showing how this endless one-way stream of people flooding into cities like Bombay or Calcutta upset the socioeconomic balance of a society. It's called *Immigrants* in English and hones in on some 40 people who share a room, the problems they face and surmount but which endlessly dog them. In my opinion, I find it to be my best film.

The Children's Film Society of India is a program of India's Ministry for Information and Broadcasting. Our agenda is to make value-based films for the children of India–which is a tall order because India is a country where we have children being raised with 17 different official languages in economic classes ranging from the wealthy to those in abject poverty. So far we have produced over 200 films–adventure films, family dramas, science fiction films plus animated features. In all of these we try to give value-based entertainment to the children.

One recent film delved into female infanticide. Through our films, we stress gender equality, and often the girl lead performs the most heroic deeds. In showing a girl achieving something, we send a message to be read between the lines by all the viewers. We also try to emphasize that a small family is a happy family and that children can cross economic barriers that traditionally have been unsurmountable in our society.

Sai Paranjpye is a filmmaker and chair of the Children's Film Society, India [24, Dr. G. Deshmukh Marg, Mumbai 400026, India, tel: (92 22) 2387-6136, 2380-2870, fax: (91 22) 2380-5610, e-mail: ncyp@bom2.vsnl.net.in].

Menakshee Shedde

Wife burning is still a common practice in India, as is the dowry system. It is customary for the family of the bride to pay the groom's family a dowry consisting of money and goods. It is commonly felt that wife burning occurs when the groom's family, unsatisfied with the dowry brought by the wife, arranges for the wife's sari to catch fire in the kitchen. Many women are killed in this way, thereby leaving the husband free to remarry–and receive a new dowry.

Because self-esteem among Indian women is low compared to Western self-perception, wives are encouraged to accept blame for the burning, often telling the police the fire was just an accident. New trends, however, have brought to light the outrageous demands some grooms and their families make through the dowry system. Recently an upper-class bride walked out of her own wedding when the groom's family demanded more dowry before the ceremony, becoming an instant celebrity and advocate for women's rights as she did so.

Menakshee Shedde, a journalist for the *Times of India*, in 1997 began writing for the FPAI newsletter of the Family Planning Association of India and has traveled widely to remote villages for her pieces – which she has found an inspiring and humbling experience. She has reported on vasectomy camps in Rajasthan where getting men to participate and take responsibility for family size is a major accomplishment. She has been to villages in Bihar where the women handle engineering tasks like repairing hand pumps while the men are out foraging. Of all the professional bylines she's had at the *Times of India*, she admits writing for the FPAI newsletter has brought her more joy.

The Real India

In Mumbai we live in a plastic bubble – with no idea of what the real India is about. Thanks to FPAI, I've gotten to see positive things happening in the countryside, which almost never makes it to the papers – that seem fascinated only with brutal murders, rapes and rubbish like that, which although true, are not the full story of what life is like in those regions.

Once I went to a village in Bihar which had no electricity or even decent drinking water – just fireflies lighting up the night. FPAI has connected with other NGOs in order to lift that village out of darkness and got funding from UNICEF to deal with the water and sanitation in order to reduce the huge number of infant deaths, mostly from diarrhea caused by drinking polluted water.

In such a village, the men worked in the forest foraging for edible tubers and roots while the women got together to learn a bit of engineering through the *mandala,* a group of ten women. They saved their money each month and then managed to get bank loans after each *mandala* decided on the amount needed to fund their goals and priorities.

Since in this region there is basically one health worker for every 5,000 villagers, these women, of necessity, must also address their own health and hygiene issues. In the village I visited, they had organized a three-day meeting where they discussed pregnancy, conception, proper diet for adults and for infants. They also talked about other issues like age at marriage and the appropriate age for becoming a mother.

Since the law on the books banning marriage before the age of 18 has never been enforced, it has had no impact in raising the age of marriage. But as the populace becomes more informed, they get involved in various programs conducted at the village level where the village elders have influence in such matters. When one proclaims, "My daughter will only marry at 18," this message starts to spread and influence others. Whereas there used to be social stigma attached to girls staying unmarried for long, now the tables are reversed because many families pressure their daughters to wait to marry until after they've turned 18, making this the new norm.

These women of the *mandala* with their savings pool, together decide how to use it for the community because the sense of community is strong among tribal groups who tend to think of the community's welfare, not just their own. Then missionary nuns taught them literacy and education which helped their health programs and income-generating skills to come together, so suddenly the entire village is lifted out of the mire it was in. Morbidity rates plummeted, as did infant mortality because women were empowered and making good decisions for their families.

All around the world, parents want their children to have a better life than they did and have more opportunities. By tapping into these desires, we can help improve the entire nation. That is why I think it is vital urban dwellers visit such villages on school outings or such, so they can envision how the real India lives (some 750 million people) and how happy it makes them to share and help one another, even in very small things.

One can discover that by volunteering to dig a little pond for a village, suddenly it has enough rainwater to last three months more so new crops can be grown and prosperity comes to the village. Pride comes to those who have participated in changing people's lives. Such teamwork makes you grow. As Mahatma Gandhi pointed out, people have a well of goodness in them that often goes untapped. It's up to us to tap it in others. I'm really proud of this country and think if people connected with the real India, they'd be terribly proud of it too. That's my vision for India.

India's Population Dilemma

India endured a horrible experience many years ago with a compulsory sterilization scheme that succeeded in turning everyone against family planning. This was a tragic setback for the family-planning program which had been progressing nicely.

The dowry system, still prevalent in India, derives from a total lack of self-esteem on the part of the woman who thinks she is worthless, so when she marries this fantastic superhero of a husband who has not bothered to get job to support himself, she feels obligated to bring him a bungalow, kitchen appliances plus a TV–among other things. Her parents will go into a lifetime of bondage to loan sharks in order to satisfy this greedy groom who is basically a lazy bum. In another few months, she'll be burned for not bringing enough dowry to keep them in luxury, and then this heroically martyred creature says, "No, no, I died while making tea." This is what comes from having no self-worth.

Why Bother?

Today few people think about population issues. It's politically incorrect. It won't get you votes or into power in the next government, so why bother? India needs to reduce its population, but few people share this sense of urgency. A sense of panic hasn't percolated deep into the country's soul–yet it's quite a nightmare scenario. We have nuclear weapons and are competing with Pakistan to wipe out half this planet, yet we can't guarantee our populace two square meals a day. Many are homeless, without basic healthcare, but we flaunt our nuclear weapons. It's ridiculous. Our environment is in a horrible state. More and more species are in danger of extinction. We're destroying our forests mercilessly and there's no vision for improving any of this. The climate's getting hotter, but no one is doing anything about it. Our population has already exceeded a billion people.

Change has to start with basic education. Then health issues must be addressed as well as income-generating skills developed. We must work together if we are to lift our whole populace out of the mire into which we are sinking. One of the easiest ways to start this is to encourage families to limit themselves to having two children. This will help us all achieve a higher quality of life.

Menakshee Shedde is a reporter at the *Times of India*, [tel: (91 11) 330-2000, www.timesofindia.com].

Mrs. Avabai Wadia, *Family Planning Association of India*

Founded in 1949, the Family Planning Association of India is the country's leading voluntary family-planning organization. Composed of volunteer members and staff, it has 40 branches and eight integrated rural projects promoting family planning as a basic human right, as well as circulating population policies. FPAI is a nonpolitical, nonsectarian and nonprofit institution.

In India, only a quarter of the health practitioners work in rural areas–the poorest and most needy communities–where 75% of the population resides. FPAI's network of over 100,000 volunteers is spread throughout the country, disseminating vital information about reproductive and sexual health and services to the most marginalized communities. Rural clinics are designed in consultation with local citizens and women's groups, allowing for greater community ownership of the clinic – as well as their ownership in seeing the clinic succeed and survive. It is the goal and reputation of these clinics that no one is turned away, even if they cannot afford services.

FPAI's well-rounded program of information, education and communication includes family life and sexuality education preparing young men and women for responsible family living, reproductive health and family planning, training and research. Its rural projects are based on local needs, conditions and priorities. Its innovative community participation approach integrates reproductive and child health with women's empowerment and youth concerns. FPAI promotes health and improves the quality of life at the grassroots level in accordance with the ICPD Program of Action. FPAI assists other NGOs in making reproductive health and family planning integral to their work. FPAI is represented on all key policy-making bodies of the government and actively participates in the formulation of national policies and programs. It is a founding member of the International Planned Parenthood Federation.

Mrs. Avabai Wadia was a founding member of FPAI and is now its president emeritus. Her career, spanning more than half a century, began when India was enduring a tremendous food shortage in the late 1940s. Food distribution centers rationed out milk and staples, but Mrs. Wadia noticed that when a mother came in with a toddler, she returned in a few years with a new baby that needed milk and food. The destitution never decreased, so Mrs. Wadia and her contemporaries began the discussion that led to the founding of the FPAI.

The Many Challenges of Family Planning in India

Family planning is still below the goals we have, since only 48% of eligible couples practice this when we would like this number to be 70%. We are most concerned about the millions of adolescents in our country who should be receiving sex education. Since schools and colleges don't teach this curriculum, we need sex education – especially in a traditional society like ours where the tradition is to obey your parents. We never encourage disobeying parents, but we do want parents and children to live harmoniously. With this terrible scourge of HIV/AIDS attacking our country, we find the tragedy is that so many young married women are getting infected who then pass this on to their children. Preventive education, of course, is the first step – so this is included in our sexuality education.

India must opt for safe motherhood. It is appalling to think that more mothers die in India each week than in the whole of Europe in a year. Maternal mortality deaths are one in 54 in India. In Western countries, they are one in 9,800. Traditionally we claim to respect motherhood, glorifying the state, so when a girl attains motherhood she achieves a new status. All that glory is useless if she is going to die in childbirth and her children are going to be orphaned.

We need a vigorous program to provide the facilities and the knowledge so women can carry out their pregnancies, childbirth and post-childbirth stages safely. At present, 34% of women receive some trained attention at childbirth. That's far too little. Of course our country is sprawled out in some 600,000 villages, some quite remote where it's not easy to provide good services. So we've started asking villagers to provide a small room in the community that is cleaned and whitewashed and made into a kind of delivery home where village women can go for primary healthcare. Here women can be attended by a trained midwife who will deliver her. Even those few first steps can make a lot of difference.

The Gag Rule

The Reagan Administration Gag Rule was first enunciated in Mexico in 1984. I happened to be the president of the Family

Planning Association of India at that time and we had to face a barrage of questions about this rule. We decided to stick to our principles, which are that individuals and couples have a choice, and where abortion has been legalized in a country, as it was in India, they have the right to choose safe abortion. The International Planned Parenthood Federation also took the stand that it cannot dictate to countries whether they should legalize abortion or not. Many countries had legalized it, and if they wished to use their own money to provide facilities for abortion, it was the decision of the national association. Therefore, IPPF gave up about $17 million in contraceptives and in cash to stick to its own principles.

Fortunately, other governments came to our rescue and in time we made up for the loss. Now the Gag Rule has come again. I believe when George W. Bush took office, he enacted this law to prevent any U.S. money going towards even discussing the pros and cons of abortion – or advising women where they could go to obtain one if they wanted it. Amazingly, it is still legal in their country for American women to get abortions, but women in poor countries are denied this aid which could benefit them from unwanted pregnancies – which is always a trauma – whether it is legal or not. This is a peculiar discrimination.

We again chose not to accept these contingency grants from the U.S. Yet, this is tragic, because family planning is now becoming universal. In the next few years, it would have been incorporated into daily health practices and we would not have had to have separate programs for that. Just as we are now confronting so many new problems of safe motherhood, safe abortion, AIDS and adolescent education, we find grants are drying up because of political tussles going on in the United States.

Donors are now either suffering from donor fatigue or are diverting their funds to other places and other causes. Grants are dwindling. In earlier years we were helped sufficiently so we could expand our programs countrywide. In 1995 we were working in nearly 10,000 villages. Now our project has been diminished so we are in only some 100 villages. Our funding was cut rather suddenly. We could have adjusted to a 10% cut per year, but we lost 50% one year with 11% the next.

From having a staff of almost 2,000 working in our rural projects and our many branch offices, we have been reduced to some 40 branches. Unfortunately the rural projects suffered the most just as we were beginning important innovations–like trying to involve the whole village community in our project. We found that when we approached them to discuss with the village what it was they felt they needed, they would invariably say they wanted a health center. Then they wanted their children to go to school and usually third on the list was a road to make them accessible or some other similar amenity to benefit the community. So we began where the discussions with the villagers had led us. We did not drop a ready-made program on them, but we did explain the importance of family planning and the effect it would have on child survival, the mother's health, in terms they understood.

Therefore we got a lot of coöperation and we located the leadership in a community. When India passed the 73rd amendment to the Constitution establishing village councils throughout the land, it turned out that in the village groups we had organized, many of those members stood for election and became the council members. These villagers, in turn, when given the chance for education, health and to be in control, took the ideas which we had been explaining to them, and conveyed them to their official councils. We felt the seeds we had planted had brought forth much fruit.

An Overview of 50 Years

In 50 years, we have accomplished certain things, but I envisage the next few years as being extremely important. This is not a time of culmination to sit back and observe what we have done, but it is an era with new problems arising that must not be neglected. We are citizens. We are involved. So long as I have my health and my faculties, I don't think I should stop working and if you really want to know, I've just entered my 90th year.

Dr. Kalpana Apte

When Kalpana Apte graduated from medical college, she wanted to help the suffering. She has never forgotten one of her first patients at FPAI who was 38, had six children but because of her tuberculosis could not be given oral pills to prevent further pregnancy. Her faith and family would not allow her to be sterilized so there was nothing she could do for her. Those woman's tears galvanized Dr. Apte's commitment to help the women of India control their fertility and improve their health

The Family Planning Association of India

When I'm out there amongst people in the communities where FPAI is working, it never fails to touch my heart that they're all part of my family and I'm a part of them. There's such a welcoming connection with these people who face hardships: women with no access to contraceptives or family-planning services, men with their own set of problems. But in spite of the difficulties they face, wherever you go people are hospitable and accepting of you.

Take for example Mrs. Chitanadan, a family-planning fieldworker working out of our Planned Parenthood Center. A great passion of her life is helping people make informed decisions about what they need to do. She goes from house to house, interacting with parents and children and is often regarded as part of the family, one who is deeply concerned with both their health-related issues and their social needs. She works in an urban slum in Mumbai where a great many households go for water and sanitation needs outside the homes, but these are happy homes.

Mrs. Chitanadan carries with her on her rounds a bag filled with contraceptives–oral pills and condoms–as well as information about immunizations, early registration of pregnancies, oral rehydration packages plus iron and folic-acid tablets for anemic mothers. If there is a first-time acceptor who might want to use oral pills for birth control, she has a checklist for the woman to make sure she doesn't have any contraindications or conditions that might not allow her to accept the contraceptive method.

Mrs. Chitanadan is one of countless workers that have enabled FPAI to realize its philosophy of reaching people to help them make choices. She's one of those building blocks that help people choose options that will change their lives.

Downsizing

A 50% cut in our budgets next year means that FPAI might have to let her go. As it is, she draws a measly salary of less than $100 per month. I do not know how we can repair this loss of trained, dedicated personnel, who can make a difference and take India to a level where communities will have replacement-size families as opposed to the situation which prevailed just a few decades ago. Then most families had seven or eight children and no one had access to the choices available today. Unfortunately the money for core programs has evaporated. It is still available for sterilizations, because the government gives grant aid for family-planning services, and is there for training NGOs for a limited number of years, but after four or five years, we have to cut staff because no one is willing to fund a core program.

Population Stabilization in India

It is definitely possible to stabilize the population – otherwise we wouldn't be doing what we are doing. The birthrate has sharply declined in India, which is not too apparent since the reproductive age population is so large. The age group at the critical junction today will not be felt for two or three decades.

An outside observer might not feel much has changed during

the last few decades, but that's not the case. In the past two decades I have observed much that has renewed my faith in the positive changes occurring all across India. Look at the progressive, rights-based changes resulting from panchayat legislation – bringing vibrant transformations in every village.

During this same period, the percentage of couples practicing family planning has sharply risen from 50% to 65% and in tribal areas from around 20% to more than 60%. These acceptors will have only two children. Knowing that health conditions indicate they will not lose any more children, they are motivated to have smaller families. This has been a true revolution, creating a miracle which has brought about profound societal changes within our own lifetime.

The Eyes of Children

It's looking into the eyes of children who otherwise have no future unless someone helps them address the problems they face, that propels me forward. We need to do so much more, yet there is much apathy to overcome. How can we make them care about and understand the issues that face our society?

In any country, children are beautiful – alive with passion and the sheer joy of being alive. They are our hope and future – whether they be Indian children, Indonesian children or Japanese children. There are those who believe in that future, and others who could care less. Yet the future of these children depends on our sensitivity to them. I love seeing children run around, happy, being a nuisance, being alive. FPAI believes in this future and wants to be there, ensuring that we make a difference in their lives.

Every child is a precious child and if this child is wanted, that completes the family. The sad truth is that these bundles of joy, with their special destinies, are often born into families that cannot look after them: mothers who are not ready to have children or want to delay or take more time to decide about the number of children they want. This is reproductive health. So we have our work cut out for us. This is the crux of the whole issue. High maternal mortality, high infant mortality, children dying before they see their fifth birthday, mothers dying during childbirth, high-risk behavior, unsafe abortions–all these are the fallouts of unwanted pregnancies.

Our country has more than a billion people in it now, yet the FPAI reaches no more than 4% of that population. "The woods are lovely, dark and deep, I have miles to go before I sleep, I have promises to keep." There is so much that needs to be done. The road ahead is not easy but it is a road that we must take. We will keep working because we believe in this–despite funding cuts, straitened budgets, a reduced staff–we maintain our belief that we are vital change agents and we will keep going on doing this.

Dr. Kalpana Apte is director of training of the Family Planning Association of India, [Bajaj Bhavan, Nariman Point, Bombay 400 021, India, tel: (91 22) 2202-9080, 2202-5174, fax: (91 22) 2202-9038, 2204-8513, e-mail: fpai@giasbm01.vnsl.net.in, www.fpaindia.com].

Sudha Tewari

Parivar Seva Sanstha (PSS), established in 1978 and affiliated with Marie Stopes International, UK, works in family planning and maternal and child healthcare. The 31 PSS clinics provide quality family planning, maternal and child healthcare services and obstetrics at affordable costs. A voluntary organization, PSS's vision has led to various outreach projects in this field.

Family Life Education enables adolescents to grow into responsible adults. Via training programs for young groups and schools, teachers and counselors provide information on issues covering the family and society, adolescence, anatomy and reproduction, interpersonal relations and planned parenthood.

To reach rural areas where the status of women is low, the quality of life poor and access to health services limited, PSS runs mobile clinics in Mewat and Gurgaon. The Mahila Mandal project uses mobile clinics to inform and educate women in their homes about reproductive healthcare and family planning through existing Mahila Mandals and youth clubs. Other promotional efforts include information and immunization camps and seminars in colleges and schools. Plans for the future include telephone information and counseling service and comprehensive and integrated family-planning services package in Orissa where awareness and service systems are inadequate.

Mahila Mandals are local women's groups based in villages throughout India. The women of the *mandala* are enabled through their collective power to advocate for women's reproductive health. Some develop their own savings and loan system. These women inspire others to come together to increase the power of their voices. Many NGOs collaborate with the Mahila Mandals to disseminate information on reproductive health. In Rajasthan many *mandala* members now delay the age at marriage of their daughters, having learned from NGOs that early childbirth is dangerous, often leading to illness or death.

The panchayat historically was a body of locally represented community leaders who gathered to make decisions and govern the legal and administrative affairs of the village. Issues ranged from providing basic and reproductive healthcare to collecting money and distributing low-interest loans based on the greatest need within the village as well as raising the level of education offered in the village.

Before Sudha Tewari became managing director of PSS, she worked in the commercial sector for eight years, building bridges and commercial buildings, learning a great deal about management. By the end of her tenure, she knew she wanted to contribute in a more substantial way to her country. At PSS she considers every day a challenge and is satisfied she's making a positive difference in India.

Indian Population Stabilization – What Does it Mean?

We believe in delivering reproductive health in an integrated, holistic manner, but our emphasis remains family planning. Population growth remains a major problem in India. Some of the southern states have succeeded in lowering their birthrates, but the northern states are still burgeoning. Unless we stabilize the population, our country will be facing serious difficulties.

The first step in population stabilization means we must empower women. Then when it comes to delivering family planning services, we must concentrate on the quality of our services if we are to ensure that women's other reproductive problems are also dealt with. For instance, if a woman comes for an abortion, we don't see her only as an abortion client, we look at other gynecological problems she may be having. Then immunize her children. We look at her as a complete woman and then deliver our abortion services. With abortion we also provide family-planning services for future reference.

We seek to provide three different strategies. The first is our clinical service which provides the entire range of reproductive health and family-planning services. Currently we have over 40 major clinics and 20 satellite clinics in Rajasthan. We work in 18 states in India.

Then we do contraceptive social marketing because we have found that in selling contraceptives to the consumers using proven commercial channels of distribution, we need to create the demand in our clients by educating, communicating and motivating them. We try to make condoms and oral contraceptive pills available and we have also introduced an emergency contraceptive product No Preg. We also handle iron folic acid because a large number of Indian women are anemic.

The third arena we address, is working with the community mostly in terms of reproductive health education and training. Here we use different approaches, such as Mumbai camps or committee-based distribution projects. We have TV commercials, TV films, documentaries and docudramas plus a large van that can travel to the villages to communicate these messages. We use many media outlets to reach the people, such as radio, cinema slides, cinema films, billboards, kiosks and wall paintings because information is the goal. Only after that we can talk about practice.

Other Dimensions of Population Stabilization

We find it is important to charge a small fee for our services if people are going to value them. When you give something for free, it's seems to be easy to slight what has been received – plus it helps us to be financially sustainable.

In India abortion is legal up to 20 weeks.

We need to enforce the law that prohibits a woman below 18 years to get married. We also need to make a girl a wanted person, not a commodity, so that everyone values a girl child. If this happens, then the need for early marriage and for dowry will disappear. It's very important we educate and empower women, but it's a long, uphill task, but I think eventually we will win.

In our Indian system called Gauna, although a girl may be married at age one, she will not go to the in-law's nor consummate the marriage unless she has a Gauna. This means she has reached puberty and has started to menstruate. We know now that though a girl starts her menses, she's not ready for childbearing, although traditionally it was assumed that if a girl menstruates she could bear a child. If we can delay the time when a girl goes to her in-law's house, this will have quite a major impact. Every day I feel satisfied that I am contributing to this country. I have an excellent team; I'm not doing it alone. There's always a challenge.

Sudha Tewari is managing director of Parivar Seva Sanstha [28 Defence Colony Market, New Delhi 110024, India, tel: (91 11) 461-7712, 461-9024, fax: (91 11) 462-0785, e-mail: pssindia@giasd01.vsnl.net.in].

S. Ramasundaram

S. Ramasundaram began his career in civil government in the mid-1980's and went to do graduate study in demography at the University of Southern California. On returning to India, he became the district collector in Mangalore in the state of Tamil Nadu. There he began lobbying for universal healthcare for all needs and this comprehensive healthcare approach was very successful and, contrary to expectations, voluntary sterilizations increased. He was awarded a MacArthur Fellowship to expand this program to 15 districts, where it was met with equal success.

Analyzing AIDS/HIV demographics for India, Mr. Ramasundaram discovered the crisis in India in the late 1990s mirrored the rates from sub-Saharan Africa in the late 1980s and so switched his emphasis from family planning to AIDS, targeting preventative research at the highest risk groups: commercial sex workers and their clients. For this he received funding from the World Bank in 1998 to replicate his Tamil Nadu AIDS Society throughout India.

The Calcutta Model

Systems are not equally efficient or equally amenable to change in all parts of the country. Tamil Nadu and Kerala, southern states, are usually more open to new ideas, but it's not impossible to replicate them in other areas.

For brothel-based sex workers around the world, I think the Calcutta model is most successful. The HIV prevalence in Calcutta has remained at around 10% for many years, but in Mumbai it quickly escalated to over 60%–making this one of the highest rates for any sex worker population in the world. This, obviously, had to be addressed. Since there were successful models available in India which could be replicated, this became the goal for Mumbai. The AIDS societies in Mumbai have begun using these models and the results will be seen shortly.

Many of the young girls in the brothels are trafficked from

other parts of India and even other countries. They get infected, go back to their villages, get married and then infect the next generation also. The Calcutta model has slowed infection levels, keeping them under control by health intervention, not by trying to rehabilitate them or branding them as evil, but informing sex workers how their current practices put their own health at risk. Changing this to safe behavior has proven very effective.

Current HIV/AIDS estimates show between 0.7% to 0.8% of India's population is infected–which converts to under four million people. Projections using different scenarios indicate increases in the range of 5 to 6% can be expected in the next ten years. So, although we have a problem, we still have a window of opportunity to control it and progress is being made in containing HIV before it reaches levels as seen in some countries of Africa. It appears India will be able to contain HIV/AIDS at much lower levels.

A Personal Context for India's Population Growth

Cities are crowded, whether Manhattan or New Delhi, because that is where people congregate–for work, for entertainment, for education. This does not cause a population problem. The population problem in India is about health rather than numbers as a whole. At the time of independence, the average life expectancy of an Indian was under 40–now it is over 60 years. These are positive signs. But at that time the average family had about six children–now it's just over three. Halving the number in 50 years is a remarkable success for a purely voluntary program.

S. Ramasundaram is joint secretary for the Department of Commerce, [tel: (01 91 11) 2301-1771, 2301-0231, ext. 466, e-mail: sundaram@ub.nic.in].

3

Indonesia

Introduction

Indonesia is an archipelago with over 3,000 inhabited islands. The Javanese are the majority with 218 million people and, except for the Hindu population on Bali, most of the country is Muslim. Under the leadership of President Soeharto, the population policy evolved from a consensus-making process within villages and urban areas to the BKKBN, the government coördinating board. All ministries that provide development services are represented on the BKKBN board. There is significant financial and administrative support by all sectors of the society, including Islamic leaders, parliamentarians, the media and the judiciary.

Over a million individuals have been mobilized to promote and deliver contraceptive services. All new contraceptive methods used in the programs have been carefully pretested and adapted to local health and cultural needs. There have been thousands of dedicated public and private individuals who have pioneered in one of the world's most successful family planning and women's reproductive health programs. Indonesia has an international training center where professionals from around the world have come to learn from this remarkable experience. What's amazing about huge bureaucracies is their ability to change and accept innovation. The program has become increasingly more decentralized.

Thanks to the pioneering efforts of Biran Affandi, chief OBGYN at the Indonesian Medical University and chair of the Indonesia OBGYN Society; Firman Lubis, director of the Raman Saleh Clinics; Does Sampoerno of the Pathfinder Fund and a host of others; extensive services have been provided by both public and private doctors. There is continuing effort to utilize all 33,000 midwives and to have all services be community based.

For many years the chairman of the BKKBN was the dynamic, creative and energetic Haryono Suyono whose tireless efforts were rooted in his great joy at working from the village up to explore innovative ways of delivering services. He did not hesitate to actively involve the president in obtaining policy support for multisectoral family planning and women's reproductive health programs. In 1993 Haryono drafted a letter which President Soeharto sent to all 108 heads of government of the nonaligned nations asking them to sign a "Statement on Population Stabilization." President Soeharto handed this statement signed by 75 heads of state to Boutros Boutros-Ghali at the United Nations in a presentation orchestrated by the UNFPA executive director, Nafis Sadik.

At the time this documentary was filmed in Indonesia, the chair of the BKKBN was Professor Dr. Yaumil Chairiah Agus Akhir who was ably assisted by Pak Lalu Sudamardi. Since the working title of the film at that time was, "Let the Women Speak," with an emphasis on quality of services, we wanted to capture visually as many of the institutions and individual players as possible, including the village family-planning management assistants (PPKBD), their supervisors (PPLKB), and village-level organizers (KADER) who distribute contraceptives and act as a liaison between the fieldworkers and the community leadership structure.

Family planning and maternal and child health are integrated within all aspects of development, which is called *posyandu*, with an emphasis on self-sufficiency. The government's welfare movement (PKK) is integral to all phases of the program, as are the primary healthcare centers (*puskentmas*). One could go on endlessly about community-based distribution centers and the myriad NGOs participating in the program, but suffice to say, there's hardly anyone in the public or private sector not involved in this issue.

This film would not have been possible without the able assistance of Rudi Pekerti who organized all the filming in the villages and mosques. Special thanks to Russ Vogel, the JHPIEGO representative who has a long and distinguished career in Indonesia dating back 20 years.

Professor Dr. Yaumil Chairiah Agus Akhir, BKKBN

Indonesia's National Family Planning Coördinating Board (BKKBN) is a government agency responsible for managing the nation's family-planning program with 27 provincial offices, 307 district offices and 53,000 fieldworkers. BKKBN relates to all ministries, nongovernmental organizations (NGOs) and community organizations or institutions. The core of the BKKBN is the family-planning fieldworker who typically can visit 10 to 15 houses per day. They not only assess contraceptive needs, but also analyze each family's socioeconomic status, reporting this information to a centralized database. The program aims to be sensitive to each village and community's needs and circumstances. In certain cases, the strategy is to create income-generating activities and the poorest households receive subsidized family-planning services.

When the BKKBN began 30 years ago, programs focused on the health of the mother and child primarily through birth spacing and postnatal programs. After the Cairo ICPD in 1994, the BKKBN broadened its outreach to include cradle-to-grave health services. It also sponsors an international training program where visitors from abroad come and observe the integrated programs that the BKKBN offers in order to improve their own programs based on their experience in Indonesia. When in-country consultation is requested, the BKKBN will send staff to the country in question.

Professor Dr. Yaumil Chairiah Agus Akhir assumed the leadership of the BKKBN after the end of the Soeharto era and passed away in 2003.

Population and Environment in Indonesia

The success of family planning did not come easily. In the beginning, over three decades ago, family planning was not accepted in this country, but now, the participation of the community is our strongest asset. We work together across the board and the language of family planning is understood by all. No one is afraid or ashamed to take part in family planning. That summarizes our success.

Our biggest current challenge is to keep a strong political commitment from the government. With over 235 million people scattered across nearly 17,000 islands having countless ethnic roots and traditions and speaking more than 300 languages, Indonesia's challenge is great – especially with a population growth rate bordering on 1.48%. Such population pressures also make protection of the ecosystems difficult. Unfortunately Indonesia started thinking about such protection a little late, so, to our regret, we now see the forests are being misused, the timber is being illegally logged and the rivers are becoming polluted. As a nation, it is better to start late rather than do nothing at all.

When you consider that our country, through family planning, has reduced the population by 80 million people in the last three decades, that translates into a decreased burden on the environment. After all, the greater the population, the more the environment will be victimized by its inhabitants. We can address all these issues positively, but we must be a little patient. Currently, our fieldworkers are doing the major work. Without them, there would be no BKKBN – and no family planning in Indonesia. If one day because of decentralization these programs should be stopped, it would be sad for the whole country.

BKKBN, International Population and Family Planning Training Center, [Graha Kencana, Jl. Permata No. 1, Halim Perdanadusuma, Jakarta, Indonesia, tel: 62-62 800-9117, 800-9029, fax: 62-21-800-9093].

Dr. Haryono Suyono, *Indonesia Peace and Prosperity Foundation*

As former chair of BKKBN, Haryono Suyono spent 35 years in the field of family planning. In 1994 he was vital to the success of the "Statement on Population Stabilization," mentioned above.

The Statement on Population Stabilization

When Robert Gillespie asked us to spearhead the crusade to get this statement signed by leaders from all over the world, especially from developing countries, in time for the 1994 Cairo ICPD it turned out to be an exhilarating experience for us. The most vital part was the participation of Indonesians themselves. It was also encouraging to interface with the efficient administration at both the highest levels and the grassroots level. It was very encouraging to have the participation of President Soeharto plus all his cabinet ministers and the governors.

The family-planning program in Indonesia started as an effort to help people utilize contraceptives under the motto we soon developed: the small, happy and prosperous family. By using contraceptives, you manage to reduce the size of the family plus the maintenance of their health. But smaller only is not enough. That is why we went beyond family planning, introducing the happiness and prosperity components–the self-help program. We supplied some support so people start their own businesses in their homes whereby women and children could be empowered. Also children were sent to school and given scholarships. Today we are still facing the economic problems that began in the late 1990s. Family planning has subsequently suffered because the government can no longer financially support it with contraceptive aid. That is why at the moment, we all have to work harder than ever.

Dr. Haryono Suyono, [Indonesia Peace and Prosperity Foundation, Gedung Gajah, Jl. Dr. Saharjo 111, Kav. Y, Jakarta 12810, Indonesia, tel: 62-21-831-9606, fax: 62-21-831-9607, e-mail: indra@ gemari.or.id; Haryono@gemari.or.id www.gemari.or.id/indra].

Biran Affandi, M.D.

Biran Affandi believes in recruiting and employing midwives as there are approximately 1,100 OBGYNs in Indonesia for a population of 235 million, whereas there are 80,000 midwives. Indonesia has a relatively high maternal mortality rate – 354 women per 100,000 live births will die, half of these deaths due to induced abortion. When a woman comes for post-abortion care in one of his clinics, she receives a contraceptive method.

Midwives and Abortion in Indonesia

The midwives in our country provide some 70% of all contraceptive services in Indonesia. From the time we begin our courses in medical school, we learn that if we are to be successful in delivering programs, we have to involve midwives. As a doctor I learned many practical skills from midwives, whereas from my professor I was only taught theoretical principles.

Officially, abortion in Indonesia is illegal according to penal laws inherited from the Napoleon Bonaparte era. In reality, of all normal pregnancies, some 10 to 15% are spontaneously aborted. With 6 million pregnancies each year, there are some 600,000 to 900,000 spontaneous abortions. Induced abortions run at about the same level because of the 26 million couples using contraceptive methods, there are between 1 and 1.6 million pregnancies which result because of some failure in the contraceptive method. About 60% of these unintended pregnancies will have induced abortions and most of these will be septic abortions – or unsafe, in non-hygienic conditions, conducted by people who are incompetent.

This means about half of the country's maternal mortality is abortion-related. For this reason we are now providing programs in our institutions for post-abortion care. What we need to do is prevent abortions in order to improve our healthcare system, but we really should avert the unintended pregnancies to begin with. To do that we have to have good contraceptive services.

Thus our first priority is supplying people with contraceptive methods. Today should a woman try to purchase a contraceptive, she may find it difficult to do so. The price of contraceptives is excessive. For example, the implant, which is very popular among Indonesian people, costs about US $44, which is very expensive for the normal consumer in Indonesia.

Biran Affandi, M.D., Ph.D., is chair of the study group on human reproduction at the University of Indonesia and also founder of Klinik Raden Saleh, [Department of Obstetrics & Gynecology, Jalan Raden Saleh 49, Jakarta 10330, Indonesia, tel: 62-21-314-0286, 315-0076, fax: 62-21-315-0558].

Dr. Firman Lubis

Dr. Firman Lubis, executive director of the Yayasan Kusuma Buana (YKB) – an urban-based foundation focusing on community reproductive health, medical services, research, training and IEC material development – began his family-planning career in the late 1960s. His areas of expertise include self-reliant reproductive health programs, community-based family planning, school-based parasite control and factory-based STD and HIV control.

During his internship at an urban hospital, he regularly saw women die from septic abortions due to a lack of adequate treatment. Even though abortion is illegal in Indonesia, scores of women faced the prospect of death, Dr. Lubis believes, due to the overwhelming poverty of large families. Often when the mother died due to a botched abortion, the family fell apart without her as the linchpin and the children often ended up on the streets. Such tragedies instilled a commitment in Dr. Lubis to work to prevent the need for illegal abortions by offering access to a variety of family-planning methods suitable to the population.

The Banjar System of Bali

When I was first involved in family planning, the focus was on the rural sector where 82% of the population lived. Traditionally, communities organized their lives – be it agriculture, irrigation or social ceremonies – around what in Bali is known as the Banjar system, a community organization that makes decisions on local matters. When people get together to decide what should happen it is known as Banjar, thus it was essential that family planning be introduced into the Banjar system. Once the Banjar opts for a program, it will most likely be accepted by the community members not only in Bali, but also throughout Java.

Model Clinics

Clinics provide different approaches to urban communities. People tend to assume residents in the cities will accept family planning easier than those in the rural areas because they are better educated. The data revealed, however, in the late 1970s that family-planning acceptance in the cities was actually less than in the rural areas. Couples living in the cities were more skeptical about family planning. Some had heard rumors or read articles claiming that taking the pill could cause cancer or that an IUD could travel in your body. Such rumors are less likely to spread in a rural community that is more homogenous. If one wants a pill, the other wants IUDs, and they want to select whatever fits their needs.

Based on these realities, a clinic was designed to meet the various needs of the people. When some complained that the government clinics didn't give out good information, efforts were made to provide as much information as possible. They also asked that clinics stay open longer and be more conveniently located to residential areas. Also nurse midwives with whom female clients felt more comfortable to speak openly were employed.

In the early 1980s when a sudden drop in global oil prices meant the Indonesian government had a reduced budget to work with, the government decided it must trim its expenditures on the family planning agency. It turned out that people were willing to share in the cost burden and the program went on. Since then the country's clinics have been used to train clinicians from Bangladesh, the Philippines, Vietnam and China who wanted to develop a similar "self-reliance family planning" whose purpose is not just cost sharing, but providing an improved quality of service.

Urbanization

Like many developing countries, Indonesia is facing rapid urbanization. With a national population growth of 2%, urban growth is at 4.7% showing the large shift from rural to urban. Every year a quarter of a million people arrive in Jakarta seeking a better life – many of these young people looking for jobs. They end up living in densely populated, lower- to middle-income areas. It's this population plus the very poor who are targeted by the YKB clinics. With over 10 million people, half of Jakarta could be described as lower income or the poor. Many of these are wanting to engage in family planning, but they want also good service, better information and opportunities for education. Most of these first come to YKB clinics when they are pregnant and either want antenatal care or are just ready to give birth.

The aim of this family-planning service is to provide a cafeteria choice – a broad range of contraceptives so people can choose the one fitting their needs. Injectables are currently very popular as are the contraceptive subdermінal implant since once it is inserted under the skin, it releases a hormone slowly for five years. The first subdermal implant consisted of six rods placed in the upper arm with a small incision under local anesthesia. Women don't have to remember to use it. It's there for five years. At one point Indonesia had more subdermal contraceptive implants than any other country. It seems this is related to a longstanding tradition in Java and other parts of Indonesia where, in the interests of beauty, gold or silver needles were inserted under the skin of a woman's face, or stainless steel needles were inserted under the skin of a macho man's arms. So when subderminal contraceptives were introduced, they were easily accepted in the country.

Population Projections in Indonesia

It was once projected that Indonesia would have over 278 million people by 2010. Thanks to family planning it seems that number is now closer to 240 million. When family-planning started here, the fertility rate was 5.7. Now this has been reduced to about 2.4 which some see as a miracle. The goal is to achieve ZPG—zero population growth—for every family, a challenge since there's a still unmet need for family planning. Another issue is reproductive infections, including sexually transmitted disease and HIV/AIDS. To combat this effectively Indonesia needs financial support from the international community or private donors because the current economic crisis precludes the nation's ability to furnish contraceptives or even to staff family-planning programs. Fortunately, a key factor in keeping family planning in Indonesia is the strong political commitment plus thousands of fieldworkers.

Indonesian Maternal and Infant Mortality

In spite of an improved infrastructure, the country still has the high maternal mortality of 350 per 100,000 births—one of the highest in the region. This means 20,000 mothers die each year in Indonesia. The technology and the methods are available, they just need to be distributed so the goal of reducing maternal mortality by at least 15,000 mothers can be achieved. The same needs to be said for the high infant mortality rate in Indonesia. By reducing this, hundreds of thousands of lives could be saved.

When I was an intern in the late 1960s, I was shocked at how poverty was affecting our country's hospitals high deathrates for mothers and infants. Daily I would encounter four or five victims of septic abortions due to the women going to quacks out of desperation and subsequently being rushed to emergency rooms where many would die. This galvanized my career choices because I realized that where there is poverty, and too many children, this pushes women into such fatal choices. At that point I committed myself to concentrate my career on family planning.

Yayasan Kusuma Buana (YKB), [Jl. Assem Baris Raya A/3, Gudang Peluru, Jakarta Selatan, PO Box 8124, Jakarta 12081, Indonesia, tel: 62-21-829-6337, fax: 62-21-831-4764, e-mail: ykb-jkt@idola.net.id].

Russ Vogel

For over 20 years Russ Vogel has worked in Indonesia for JHPIEGO, a not-for-profit international public health organization affiliated with Johns Hopkins University in Baltimore, Maryland. This agency has been working to improve the health of women and their families throughout Africa, Asia, the Middle East, Europe, Latin America and the Caribbean. JHPIEGO's work includes prevention treatment in reproductive health and family planning, HIV/AIDS, maternal and neonatal health and cervical cancer.

After President Soeharto resigned, many government organizations providing health services were decentralized. JHPIEGO has collaborated with the government of Indonesia to create and expand access to services. A lot of people realized the progress they had made could be lost, so they began thinking not so much about moving ahead, but just sustaining and helping the government maintain what had been achieved–which was at risk with the decentralization, the economic crisis and the social upheaval making everyone afraid of slipping backwards.

The Key Is Partnership

JHPIEGO came to work not only with NGOs, but also to coöperate with the government officials who, after all, are the ones with the authority and the most access to money. That's why I emphasize the sharing partnership they have with us. Mutual decision-making is the key, not just experts coming in and telling them what to do. This is a vital stance for donors to take, wherever they work, whether it's in Indonesia or elsewhere.

One of our main jobs is advocating at the district level for family planning–just as in the U.S. we need to advocate with our state governments–because this is the level where the family-planning programs, child health programs or programs for the poor are in place. This is a change from a few years ago when everything was mandated down from the central bureaucracy. This means reaching out to all 370 districts and cities if they want to make sure these programs are maintained at the district levels. But these programs must compete for support with roads, agriculture, education and health projects who are all trying to make sure the district governments funds their programs as well.

There is dedication and there are good ideas–plus a real attempt to make the family-planning programs serve the needs of the people in the face of competing requirements from all different levels. With the new democracy in Indonesia, you have a major impact of politics, but politics means compromising. You have to have politics when you have democracy. We have to live with it.

JHPIEGO/Indonesia, [TIFA Building, Suite 1002, Jalan Kuningan Barat #26, Jakarta 12710 Indonesia, tel: 62.21.520.1004, fax: 62.21.520.0232, e-mail: jhpiego@jhpiego.or.id].

Bob Gillespie on Indonesia

The Family Planning Fieldworkers

The family planning fieldworkers in Indonesia have been providing reproductive health services for 35 years with a very comprehensive approach towards caring for women, increasing the age at marriage, caring for each child as that child develops in that particular neighborhood, visiting every single household twice a year and allowing a depot distributor to give out condoms and pills on a constant basis. All of this combined is what makes a family planning program work–the door-to-door fieldworkers.

These same activities were taking place in Korea and Taiwan in the mid-1960s, so this basic formula has been effectively combining women's reproductive health with family planning for nearly four decades in some parts of the world. What's exciting is that year in and year out the level of commitment to the program has been sustained. Women and men have been marrying at a later age. They've been spacing their births after marriage. They've been increasing the spacing between births and ultimately having replacement size families across much of Indonesia.

4

Mexico

Introduction

Mexico has 106 million people. The estimated total fertility rate varies from the official government figure of 2.4 to the United Nations Population Division figure of 2.8. Three-quarters of the population lives in urban areas. Between 300,000 to 350,000 immigrants move to the United States annually.

We interviewed the secretary general of the National Population Council (CONEPO), Elena Zuniga Herrera. CONEPO is responsible for coördinating policy, mandating laws, conducting research, developing family-planning/women's reproductive health guidelines and drawing on all the government ministries to provide one of the most expansive and comprehensive family-planning programs in the world. To determine the extent of the services provided by the Ministry of Health, we interviewed Dr. Maria de Lourdes Quintanilla Rodriguez.

Mexico is unique since 42% of all workers are covered under a social security system served by an extensive network of clinics, hospitals, trained midwives and community-based family-planning fieldworkers. This also includes an education program for adolescents. The director general of the International Planned Parenthood Federation affiliate (MEXFAM), Alfonzo López-Juárez was most helpful, arranging interviews with government officials, setting up filming sites in rural and urban areas plus giving us access to many in both the government and private-sector programs. Dr. Ricardo Vernon of the Population Council was helpful in discussing important research activities focused on delivering family planning and health services.

We filmed traditional and trained midwives, *promotoras* (fieldworkers) and village leaders to see how family planning and women's reproductive health services were provided by the government and private sector and were impressed by the successful adolescent programs under the auspices of the social security system and MEXFAM'S *Gente Joven* (Young Adults) campaign. In both urban and rural settings we were able to capture visually the mobilization of private doctors and the social marketing campaign, interviewing over 100 individuals ranging in age from eight to over 80.

The women in villages are given a voice to express their views on the comprehensive health and family-planning services. In Mexico one can see the contrast between young teenagers learning about sexuality and grandmothers voicing their opinions on the radical transformation that has taken place in just two generations. We also explored the issues of AIDS prevalence, environmental concerns and the impact of NAFTA on the farming community.

Elena Zuniga Herrera, CONEPO

Consejo Nacional de Población (CONEPO) was created to coördinate the population policy for Mexico in 1974 during a period of rapid population growth. A law mandates that public education provide population and family-planning instruction and also states that family planning must be provided free of charge in all public health institutions. CONEPO is composed of 12 agencies from the federal government, including the ministries of the interior, foreign affairs, social development, education, labor, agriculture, agrarian reform, and health, as well as the IMSS and the Social Security Institute for Government Workers. In 1983 CONEPO created state population councils and its programs evolved using information collected through national surveys and population censuses. Their research covers immigration, emigration, fertility, demographic projections and social indicators.

From Federal to State Population Councils

In 1993 state population councils were created to coördinate annual population figures, family planning and women's health plans. Every six years CONEPO develops a plan which the state population councils adapt to fit their state realities or needs. Although population policy is formulated at the federal level, it is the state population councils which adapt this program to meet local needs. In 1992 some 894 municipal population councils were created. CONEPO develops plans based on population censuses and national surveys that research migration, international migration, urbanization, fertility and projection of demographic indicators. A special concentration is on young people and rural women – as well as on the aging population. Population and environment studies or surveys are complex as they deal with varied regions – coastal areas, jungles, those locales where people are facing challenges or issues with ecology. But these studies also encompass the effects of the demographic pressures such as water and infrastructure.

In the 1960s the net loss of population from Mexico to the U.S. was 30,000 people a year. That figure today is estimated by some to be closer to 380,000 people annually. And where migrants originally came from rural areas, today they tend to leave urban areas. This means the number of municipalities in the country contributing to this migration has increased. Thus Mexico is losing the working age population, an age where young people are the most productive and there are social costs from the growing separation of families, of parents and children. Of course there are benefits from having migrants who are successful and help their families achieve a better standard of living in Mexico. Current information indicates some 10 billion dollars were transferred last year from the U.S. to Mexico. Due to NAFTA, the economic geography of the country has changed with migratory flows increasing. The rural exodus is high especially among the working-age population, so small farmers who can no longer compete with food imported from the U.S., especially wheat and corn, are moving into the cities or across the border to the U.S.

The total fertility rate in Mexico is currently 2.3 children per woman. In the 1970s when this policy was launched, the general fertility rate was 7.4 children per woman. In some areas the fertility rate is still five children per woman. For young people the problem has gotten worse. In rural areas the onset of motherhood and first pregnancy is much earlier. Few wait until age 22 or 23 to have their first child. In urban areas, an increasing number of women have their first child out of wedlock or at a very early age. Most teenage girls initiate their sexual life at 17. Unfortunately this young group has a time lapse between the onset of sexual activity and the birth of the first child of only 11 months.

Over half of Mexico's inhabitants live below the poverty level – which is classified in three categories. Those in "extreme" poverty, 17 million people, have the highest rates of mortality, highest rates of fertility, the lowest prevalence of using contraceptives, and an earlier age of marriage. This population is also the one with the highest unmet demand for contraceptive methods. It is estimated that if fertility rates in the 1970s had not changed, the population in Mexico would be 160 million instead of the 106 million today.

Consejo Nacional de Población, [Ángel Urraza 1137, Col. Del Valle 03100 México D.F., www.conapo.gob.mx, tel: (52) 54888401, 02].

Maria De Lourdes P. Quintanilla Rodriguez, *Ministry of Health*

Consejo Mexico's Ministry of Health supports programs in 17,000 health outlets, including 2,000 hospitals. Services begin with prenatal care, childbirth and postnatal care up to the age of two throughout the entire reproductive life cycle. Approximately 12 million women use contraceptive methods supplied through the MOH. Currently, the ratio of tubal occlusions to vasectomies is 18:1. The gap for permanent sterilization has closed from the early 1990s, when the ratio was 56:1. To promote gender equality, the health ministry developed a program for male involvement called New Masculinities. Each day 1,100 teenage girls give birth in Mexico, but daily two teenage mothers die. The MOH promotes the delay of sexual activity onset and reinforces sexual responsibility by targeting college students, non-students, gang members and at-risk adolescents. Over 30,000 traditional midwives are getting certification and training to incorporate family-planning services.

A Comprehensive Program

Every state has its own Ministry of Health hospitals which work with the social security institute for government workers and with the Mexican oil company, Pamex, not only on family planning, but also in the field of reproductive health for males and females. The national congress and the chamber of deputies and senators are interested in supporting these programs. An emphasis has been made to link health education to youth sexuality, both with college students and non-students. *Machiladora* workers along the border tend to be young people who initiate their sexual activity at an early age, so efforts are being made to encourage the delay of the onset of sexual activity among this population.

Currently there is the painful situation facing all these programs because of a withdrawal of resources. Consequently supplies are dwindling because in the past these were funded by donor agencies. Now state governments themselves have to provide them. The joint purchase along with UNFPA of contraceptive supplies has helped. The role of NGOs has been crucial, not only in terms of creating awareness about family planning, but in helping to improve the quality of services provided and developing innovative programs. Traditional midwives deliver 370,000 births each year. It is vital they be encouraged to be part of the solution because currently not all of them do family-planning counseling. By providing training to them, our family planning and maternal and child health services program could be much improved.

Doctors and nurses are being trained to use folic acid, iron and micronutrients with their patients to help children get a good start in life. This is a special need with the indigenous populations and to help meet this, training materials in twelve different languages have been produced.

C. Maria de Lourdes Patricia Quintanilla Rodriguez, Centro Nacional de Equidad de Genero & Salud Reproductiva, [Homero No. 213, 6° piso, Col. Chapultepec Morales Deleg., Miguel Hidalgo D.F., C.P. 11570, Mexico, tel: (52 55) 5545-3733, fax:(52 55) 5545-4558, e-mail: lquintanilla@salud.gob.mx. Ministerio de Salud de México, Lieja 7, Col. Juárez, 06696, México D.F. www.ssa.gob.mx].

Dr. Javier Cabral Soto, Programma IMSS Oportunidades

The Mexican social security institute (IMSS – Instituto Mexicano de Seguridad Social) reaches 45 million people, focusing on states with high levels of poverty. The infrastructure consists of 3,500 medical units and 69 hospitals linked by radio to extensive health services. The medical units have doctors, community promoters (*promotoras*) plus the participation of some 4,000 rural midwives who receive training, are provided with supplies and create a crucial link between modern medicine and traditional rural medicine.

Reaching the Youth of Mexico

IMSS developed a program targeted at rural youth, CARA – rural centers for adolescents – because maternal and infant mortality rates are twice as high in rural areas as compared to urban rates, partly due to early onset of childbirth and quick succession of pregnancies. CARA, operating through the IMSS medical units, reaches two million rural youth. About one million youth now practice contraceptive methods. Each CARA offers education and services for youth aged 11 to 19. In addition, rural youth can lobby for a CARA outlet and receive training and supplies to establish CARA in their village.

Eight million Mexicans are given health services as a direct result of contributions made by government workers. Some 60% of the women in the reproductive-age group are provided family-planning services and our program operates in 17 of our 31 states. There are 46 native languages spoken in Mexico and the rural midwives, who live in their own communities and speak their own dialects, provide a valuable linkage between modern medicine and traditional medicine. Primary healthcare physicians are located in the health centers.

For eight years we have been working with young people with high fertility rates and the use of contraceptive methods among young people has doubled through our centers during this period. This program is focused on rural areas where the culture may be different from that in urban locales where more accept the use of modern medicine. The success of this rural endeavor has dispelled the myth that the rural poor resist participating in such programs.

Even though there is a high usage of our services in urban areas, there is still some reluctance in rural areas, but we are persisting with programs of education and communication. To be successful we always encourage these rural communities to partner with us so we never go to a locale without having reached an agreement with them to work together. Right now in Mexico we have a project linking health, education and nourishment. The name used to be "Progress;" now it's called "Opportunities."

I am often invited to make presentations or lecture at conferences describing our program in Mexico. I always turn them down, suggesting rather the best way for people to learn about our program is to come here and work alongside our people in these rural areas so they can learn from our experience what is happening here. That's also why we have produced and written materials and books that can be copied, replicated else where.

IMSS, [Av. Paseo de la Reforma 476, 3er. Piso, Col. Juárez, C.P. 66698, D.F. México, tel: (52-55)5238-2700, www.imss.gob.mx].

Alfonso López-Juárez, MEXFAM

The Mexican family-planning association (MEXFAM) is a non-profit association specializing in offering services to the poorest sector of the Mexican population. Founded in 1965, MEXFAM is a member of the International Planned Parenthood Federation (IPPF) with a mission to provide quality, innovative services in family planning, health and sexual education. MEXFAM works in both cities and rural areas and its services reach more than 400,000 families each year.

Gente Joven is a program targeting 1,600,000 young people between the ages of 10 and 23. Counselors, recruited from this age group, are trained in aspects of sexual responsibility and health and then sent to engage in peer-group counseling. *Gente Joven* provides information at an early age, helping youth to delay the onset of childbirth and pursue their educational dreams, breaking the cycle of poverty. The social security system uses the *Gente Joven* program to reach into poverty-stricken rural areas and conduct sexuality fairs in public schools, replete with free condoms and games on sexual responsibility.

40 Years of Success

MEXFAM was among the pioneer organizations begun in the 1960s to work on population issues, but now it is the only one left in the field. During these 40 years much progress has been achieved in Mexico. While in 1965 there was virtually no use of family planning in the country, latest statistics claim 70% of couples here now use contraceptives.

The big problem in Mexico is that although young people use contraceptives once they have two or three children, they begin having children when they are 16 or 17 years old. Once these young people become parents, they have to end their education and take low-paying jobs to support their family and are stuck in a cycle of poverty. The second problem is that 30 million people live in remote areas and have no access to family-planning services.

To address these issues, MEXFAM implemented *Gente Joven,* described above, and also community-based programs in remote, rural areas where health services of all kinds are offered. So far 30 reproductive health clinics have been established where myriad services are offered. We also franchise family-planning activities in small, private clinics which produces income to subsidize services for the poor and we produce instructional materials.

Our *Gente Joven* program has been replicated by the social security system that is working in 17 of the poorer states in Mexico and in each one of their clinics, there is a special program for young people. One important outlet to provide contraceptives for young people is the pharmacist. Traditionally pharmacists were not seen as friendly to the young people, so we started a program called Youth Friendly Pharmacists. The success has been significant in helping young people find the contraceptives they need. In Mexico, almost all drugs are sold over the counter, so that is advantageous for our program.

One issue we cannot avoid is the sheer number of people in Mexico. In comparing Mexico to its neighbors to the south, Guatemala has a tenth of our population and all Central America together has perhaps a third of our population. The volume of people in Mexico means that even though our growth rate is less than that in Guatemala, still the growth produced each year is huge – almost two million people annually.

The distribution of wealth in Mexico is quite unequal, so

where the middle and working classes in Mexico have no problem accessing family planning because social security provides this to them, half of the Mexican population is under the poverty line with no access to health services, or any kind of services.

The population growth rate in the U.S. is higher than in Britain and in Germany because the use of contraceptives is not as open or as common there as in Europe. Americans like to consider themselves as very advanced, and in many aspects that is true, but in regards to sexuality, they are not at the forefront. There are segments that continue to insist on abstinence programs while research shows there is no evidence they have any effect. Statistics show young Americans are as sexually active as their European counterparts, but they do not have the same access to information or medical services. As a result, the number of unwanted pregnancies in the U.S. is high. This is even more evident in the American Latino population and the African-American population – where almost 60% of new cases of sexually transmitted diseases are found.

If we in Mexico continue on our current path, we can hope to stabilize our population by 2050. A huge challenge we cannot ignore is that our populations are blending north of our border. In the U.S. 12% of the population is of Latino origin and 60% of that population is Mexican origin. The growth rate of the Latino population is much higher than that of others, so in the future, more Mexicans will be living in the U.S. It would behoove both countries to find a solution to the current migration issue. Since our geography and history have united us forever, that should give us a good basis for helping each other wherever we can and then look to give both countries, which are going to have mixed-race populations in the future, find solutions to our common problems.

The Mexican Family Planning Association (MEXFAM), [Juárez 208, Tlalpan 14000, D.F., Mexico, tel: (52 55) 5487-0030, fax: (52 55) 5487-0042, www.mexfam.org.mx].

A MEXFAM Sex Education Class

The following is the text of a speech given in a public middle school classroom:

AIDS attacks our immune system, and that's why we call it Acquired Immune Deficiency. AIDS attacks white blood cells. There are five places where we have these white blood cells. In blood, of course. Where else? We have white blood cells in blood, in semen and pre-ejaculatory liquid, sexual bodily fluids in the case of females and in maternal milk. In these five places we have white blood cells, and that's where the virus can live and develop. Now, how does the body absorb it? Through the mucous membranes. We have these kind of membranes in our body in four different places: the vagina and the vulva, which is the external part; in the penis; in the mouth; and in the anus. So any practice involving white blood cells and these mucous membranes leads to a risk of infection. That's why contact with the penis to the mouth or contact of the vulva to the mouth, even though there is no blood, risks infection.

So now we'll talk about the use of condoms and I will need a volunteer. We have our own model. I mean, males never participate, so we have a model and our model is called Pancho. The condom should be rolled down when the penis is erect. You should remove it while the penis is erect. After an ejaculation, the penis still remains erect for a few seconds and that's when you have to remove the condom. If we remove it after that, the penis becomes flaccid and liquids can spill out. So while it is still erect, remove with your fingertips. Once it's removed, tie it and throw it away. We have to tie a knot to make sure the condom is not used again.

For every sexual intercourse you need to use a new condom. Some of you have begun having sexual intercourse and you don't even know how to protect yourself against unwanted pregnancies or STIs, so take advantage of this opportunity. This is not available to all schools. You have a demonstration, you have information, so please don't feel embarrassed. Ask any and all the questions you have. These are personal questions, but it is necessary to address them. We are experts and won't mislead you. On the contrary we will help you clear up your doubts.

Contraceptive foam is another option. There are several spermicides, jellies and foams that we can use. The only contraceptive method for males is the condom. The problem here in Mexico is that people believe it's only up to women to take the responsibility of contraception. It has to do with biological reasons. It's easier to control the biology of women as opposed to males. The important thing here is that in a couple's relationship, the two partners should decide as to the method they want to use and then take care of each other.

Visit with a Traditional Midwife in a Rural Village

A midwife was counseling her patient on contraceptive methods and addressing the myth of lactational menorrhea, telling her it was not true that breastfeeding women don't get pregnant to make her realize she needed an option. She discussed the IUD as a possible method and told her that she had information and that it's easy to get because they have a health center here where it can be done. For the time being she will get some condoms for her patient. The midwife suggested that her patient talk about contraceptive options with her husband so he can be also involved and know what it is all about. Finally the midwife described how the IUD works, how it can be inserted, when the patient has to get it, how long it will work, along with other information.

Javier Dominguez Del Olmo, United Nations Population Fund

UNFPA supports the Mexican government in its efforts to achieve reproductive health rights, specifically at the state level, by developing a model of intervention that increases the quality of reproductive health services provided mainly to rural and indigenous populations in the ten states selected through indicators designed by CONEPO. High-risk groups include adolescents, especially young women with high reproductive health risks.

UNFPA helps these states strengthen their ability to offer services according to government regulations. As procurement of contraceptives is not currently meeting the population's needs, UNFPA has helped the ministry of health with technical coöperation to buy contraceptives at lower prices. This permits the government to free resources needed for other activities. UNFPA supports sex education within public schools. One initiative trains teachers to share information about different reproductive health issues with children and their parents.

The Life Cycle of Sexual Education

We need to understand that sexuality is part of our humanity. So you begin with children in school and provide them information about the importance of sexuality and the responsibility that it includes. Also, it is important young people know how to protect themselves against HIV, against pregnancy, against sexually transmitted diseases and against violence. Including this education in the school curriculum is an important strategy to create responsibility among the students. Talking with the parents about this kind of education is sometimes part of that education. They also learn, because many parents don't know how to talk about sex with their children and through these programs, they learn how to deal with these issues within the family circle. It is important to introduce sex education in the schools of Latin America.

I'm a family medicine doctor and deal with this issue daily. I feel by working in this area I am helping people in the most intimate issues of their lives. I talk about sexuality, reproduction, when to have children, how to protect them and how to teach them to avoid risks related to reproductive and sexual issues. I am comfortable in my efforts because I am certain that when you improve the quality of reproductive health services, you are making a contribution to the quality of their lives and also to a full exercise of reproductive and human rights.

UNFPA México, [Apartado Postal 105-39, México City 11591 México, fax: (52 55) 5254-7235, 5255-0095, www.unfpa.org.mx, e-mail: unfpa@unfpa.org.mx].

Ricardo Vernon, Population Council

In Mexico the Population Council has two main areas of work: Operations research applied within the health system strives to achieve goals more cost efficiently and provide better quality services. Projects have been conducted on the attitudes and practices of adolescents as well as vasectomy acceptance. Data is collected to better assess obstacles that still exist and solutions to overcome them. Recent research has centered on how to prevent maternal deaths and increase access to emergency contraception.

Institutionalizing Family Planning

Family planning is one of the most publicly supported services. Many see family planning and contraception as synonymous with freedom in our country. It's used as a platform and incorporated into all sorts of legislation. The general education law says the public education system has to include information about the benefits of family planning. Our general population law says family planning has to be provided free of charge in all public health institutions and the general health law talks about family planning. Family planning now is very institutionalized. Recently we've been working on emergency contraception, in post-abortion care and in research on how to prevent maternal deaths.

Population Council, [Apartado Postal 12-152, C.P. 04021, México, tel: (52 55) 5999-8630, fax: (52 55) 5554-1226, 5999-867-3].

5

Ghana

Introduction

In New York we interviewed Dr. Awoonor-Williams, district director-general of medical services for the Nkwanta Health Development Centre; Dr. Frank Nyonator, director of the program and monitoring-evaluation division in the Ghana Health Service; and Sam Adjei, deputy director of the Ghana Health Service. Each interview concentrated on the national population/family-planning program and the role of the Navrongo experiment in transforming the health and family-planning program in Ghana. We also interviewed Jim Phillips, senior associate in the policy division of the Population Council. Most of the questions and responses evolved from the papers, "The Impact of the Navrongo Project on Contraceptive Knowledge and Use, Reproductive Preferences and Fertility" and "The Influence of Traditional Religion on Fertility Regulation among the Kassena-Nankana of Northern Ghana," published in *Family Planning Studies*. Accompanying the health professionals was a nurse-practitioner, Rofina Asuru, responsible for training health aides in the community health centers in Navrongo.

Both the interviews and the film footage portray the activities of the community healthcare centers, the volunteers, the social marketing and how in rural communities large gatherings are used to promote health and family planning with the support of local chiefs and elders. There was a long interview with Frederick T. Sai, a world-renowned expert in health and family planning who distinguished himself as chair of the United Nations Population Conference in Mexico City. He also has been a technical consultant in reproductive health at the World Bank and was past-president of the International Planned Parenthood Federation. Now Dr. Sai is principal consultant to the ministry of health on family planning, reproductive health and AIDS prevention in Ghana.

Navrongo Health Research Centre

In 1987 the Ghana Ministry of Health founded the Navrongo Health Research Centre (NHRC) to investigate major causes of morbidity and mortality in the most impoverished and isolated region of the country. NHRC'S initial project was to assess the impact of vitamin A supplements on childhood mortality. The success of this study led to a micronutrient policy and to the commitment of regional and international health agencies in providing vitamin A supplements nationally. After this vitamin A trial, the ministry of health established the NHRC as a semiautonomous research agency composed of independent scientific teams to conduct research programs in collaboration with the Ghana Health Service.

In 1992 NHRC established a demographic surveillance system designed to be portable and transferable to other health research projects. In the following years this system has served as a platform for all NHRC research, providing accurate and comprehensive accounting of all demographic events and population characteristics. Beginning in 1994 international exchanges were launched that fostered the transfer of the Navrongo Demographic Surveillance System to other projects in Africa and Asia. Sites involved in this work now share a common platform for research that supports vaccine trials, health policy projects and sociodemographic studies in 16 sites in Asia and Africa.

Navrongo Health Research Centre, Ghana Health Service, [Box 114, Navrongo, Upper East Region, Ghana, tel: (233 742) 22310, 22380, fax: (233) 21 22320, www.navrongo.com, e-mail: ahodgson@navrongo.mimcom.net, www.ghana-chps.org].

Dr. Frederick T. Sai

Dr. Frederick T. Sai, world-renowned expert in health and family planning, is now a presidential advisor on reproductive health and AIDS prevention in Ghana. He began his career as a physiologist in 1954, studying nutritional practices in the British Empire in Africa which showed a strong link between childhood malnutrition and rapid fertility. He was one of the founders of the Ghana Planned Parenthood Association.

The Task Ahead

What keeps me motivated is simply the fact that the task is not done. Women are still dying during pregnancy and childbirth or from unsafe abortions and there are still many women and couples who do not have access to family planning.

African leaders, until the ICPD in 1994, had expressed concern by how some spoke of population issues–the African populations were growing too rapidly, thus interfering with development. But humanity was absent from the discussion. The ICPD helped African leaders recognize that fertility regulation and infant mortality, child and maternal mortality were interrelated. This encouraged leaders to talk about how to improve the lot of women and their communities by including specific population-related initiatives.

The Navrongo program in Ghana is very successful. Moreover, the Ghanaians have endeavored to see that the lessons are applied as soon as they are learned. Most importantly, people aren't being told, "Come at 8:00 in the morning and you will be seen." Rather, the program is tailored to people's schedules and needs. A woman may be cooking and if she decides she wants to stop and go see the nurse midwife, she goes and sees her. The other aspect of this is that the people's own beliefs about health and illness and what is permitted and not permitted in terms of their religious practices and beliefs about spirits and ancestors are taken into consideration far more sensitively than has happened in many programs I have seen. Navrongo has shown how we involve people in their own healthcare at reasonable costs.

Every woman should be assisted in any way possible if they want to work. Gender violence must be dealt with to help women begin to feel confident about their security. In some of our communities, a woman cannot go to the hospital by herself; a man has to be there. A pregnant woman has to have her husband's permission before she goes. If the husband is not there, what happens? These are things that need to be rectified.

Some communities have arranged with truck drivers so if a card is given to them, they will take the woman to the nearest hospital or clinic for assistance without charging a fee. Death can occur very rapidly and so transportation is crucial. Requisite equipment and trained personnel at that service point is very important – antibiotics, transfusion fluids or even blood if necessary. Women can also die due to the aftereffects of unsafe abortions. Whether from a natural miscarriage or an attempted abortion, complications can kill the mother.

In some countries 30% of women who die from pregnancy-related causes, die from unsafe abortions. We have laws covering abortion that need to be looked at, but whatever happens, doctors should be trained and equipped to handle unsafe abortion without being judgmental. We are not the custodians of society's morality; we are custodians of society's health. We have to be moral, yes, and we must be examples, but we shouldn't judge people at a time when they need our services.

We ought to train our doctors to be able to handle any situation, including infection or bleeding. We have got policies regarding sexually transmitted diseases. We need to train husbands. Female genital mutilation relates to education. Ghana has a law criminalizing female genital mutilation, so some women are running to neighboring countries to get it done.

Ghana Health Service

Ghana Health Service provides a policy framework for implementing health services as well as handling issues of evaluation, fund-raising and financial distribution and providing human resource needs for the subdistrict, district, regional and national sectors. As less than half the population has access to skilled birth attendants at the time of delivery, the government recruited and trained 12,000 traditional birth attendants in basic skills of hygiene, improved delivery skills and prenatal examinations that diagnose problems needing referrals to trained midwives. The community-based health planning and services initiative relocates trained nurses in underserved communities to work in collaboration with volunteers providing door-to-door services.

Dr. Frank Nyonator, Ghana Health Service

Dr. Nyonator's division has three primary functions: looking at policies that drive the service provisions; evaluating plans and budgets; and evaluating the delivery of services.

Blueprints for a New Initiative

My division looks at policies that drive service provision in the Ghana Health Service. This includes the plans and budgets that we can put together for the health service and then we also monitor and evaluate the processes by which we deliver services. Our routine monitoring system captures health data from the district level to the regional level to the national level.

Malaria is a key health problem in Ghana and accounts for about 45% of out-patient attendance in our facilities. We also have a whole range of preventable diseases, intestinal worms and diarrheas that we have to deal with. We promote condoms and in the more enlightened communities we see a lot more injectable use rather than oral contraceptives.

The health staff used to invite people to be volunteers, but now volunteers are selected by the community. There's huge social capital in our communities that combines their resources for the good of the community. One problem is that if you send somebody who is not competent, the community's support for these initiatives will collapse, so we have set up a monitoring system at our headquarters where we get quarterly returns from the districts which clearly indicate how they are progressing over a 90-day period, and we put that together in a database and try to project it in a picture way to back to the districts.

It's an exciting moment for us all. It's an initiative that gives us a lot of challenges to work on. We have been thinking about how to provide services to the clients that need it most. Since 40% to 60% of our population does not have access to healthcare, we believe by putting this process in place, we will at least get over 70% to 80% of our population some basic healthcare.

Dr. Sam Adjei, Ghana Health Service

Dr. Sam Adjei is responsible for policy direction of the health sector and providing technical direction for healthcare delivery. His own mother, who had ten children, saw the effects of rapid population growth on Ghana and advised him to have, at most, three children.

Challenges for Ghana

In the public health sector there are around 1,000 physicians, yet there are more Ghanaian physicians in New York than there are in Ghana. Service is severely compromised because we have two medical schools that turn out around a 100 doctors per year and quite a lot of them emigrate. We don't have enough doctors to go around. Less than half of the population has skilled attendants at the time of delivery. Over 70% of rural births are delivered by traditional birth attendants.

Ghana Health Service is responsible for insuring that family-planning services become available and are utilized across the country. We also make sure that midwifery practices are safe and deal with reproductive health diseases that commonly affect women, infections, bleeding problems, complications of pregnancy and now HIV/AIDS.

The public health department recruited and trained nearly 12,000 traditional birth attendants in basic skills of hygiene and improving their skills in delivery, but more importantly to recognize problems and refer them to trained midwives. When it comes to emergency obstetric conditions, which often cannot be predicted, the traditional birth attendant is very limited in what she can do. For example they cannot give antibiotics or blood transfusion or deal with extensive bleeding during delivery. Still, they have a role in advice and education.

We work with a large group of private physicians to make up for the gap in service delivery because the public sector cannot reach everywhere. Even though most private practitioners–doctors, midwives and pharmacists–are centered in urban areas, they cover nearly 40% of the service delivery and we give them support in immunizing children. Our registered midwives provide extensive coverage in terms of delivery. We've improved the quality of services by engaging male clients and by learning how to partner with community governance systems that choose the elders, the traditional systems and the gatekeepers in the communities. More importantly we've learned how to finance systems with communities' existing financial mechanisms.

If we were to scale up rapidly, we would need over 5,000 new nurses to be deployed in communities–plus the whole package of logistics: vehicles, transport to go from house to house, motor bikes, bicycles, drugs, medications, family-planning contraceptives, etc. Then you need to find housing for nurses. It costs a minimum of $5,000 to construct a community health compound. If you want to offer immunization, you need "fridges" to store vaccinations, syringes, needles–all costly items in our economy.

I sometimes wonder why I am still in Ghana. When I was in the U.K. recently I learned they are looking for 5,000 physicians annually for the next five years. The easiest way is to go out and recruit. It's attractive when you pay Ghanaians more and they emigrate to improve their lifestyle. This leaves us at a disadvantage, since in many ways we cannot compete with the health system in the U.S. or in the U.K. But if we offer enough to make our health workers comfortable, they will not leave.

When we talk to donor communities. We found it is easy to get them to pay for the hardware of healthcare: medicines, cars, motor bikes. In some cases they've even gotten helicopters to transfer people to provide services, boats and other things, but nobody wants to give Ghanaian healthcare workers $100 per month to keep them working here. Nobody is interested in something that mundane.

HIV/AIDS is a major problem. There is quite a lot of trade and long distance travel across the borders. Epidemic diseases do not respect borders so you have yellow fever, cerebral spinal meningitis plus other such epidemic conditions crossing borders. Fortunately, our former president recognized that AIDS was a potential problem and gave political leadership to the issue, so we have an extensive educational program. But people are still concerned that if you promote condoms, you encourage people to get involved in sex. Evidence doesn't show that. It's like somebody saying that if you go out with an umbrella that means you are causing the rain to fall, but you are just protecting yourself. Where we have the biggest challenge is with treatment, because soon there are going to be more cases. Even though the prices of antiviral drugs have come down to about $300 per person, it is still a big challenge to the health system to be able to pay for such treatment.

We are hoping to get women to participate in mental development as well. This has been one of our biggest problems. We have equal enrollment of boys and girls at the beginning of the educational cycle, but after ten years of education, girls tend to drop out, so the proportion of women getting a higher education is drastically reduced. We are trying to keep girls in schools so they can achieve a higher level of education.

We depend a lot on routinely collected information to tell us what our needs are and the pattern of diseases commonly afflicting our people, such as malaria, HIV, reproductive problems and malnutrition. By looking at these statistics, we can make decisions about what is technically feasible, cost-effective, will provide major benefits when addressed, and what will be the target age group that is most afflicted. We also have to look at certain cost-cutting issues that affect the delivery of service to the target people and decide what the issues are about quality of services, their efficiency and how we could partner with others in the health sector. By considering these factors and our priorities and then discussing them with colleagues, donors, communities and politicians, we develop our strategic plans for the future.

The health center must take a holistic approach that includes family planning, population services and health services – addressing what people want and how they want it organized. It is easy to sit in the capitol, but it is more difficult to enter the community and use them as your consultants so they can advise you on how *they* do things, what *they* want and how *they* want it.

I realized that if my mother was willing to adapt, you can get anybody to change. My mother, who had ten children, urged me not to have more than three children. When I asked her why, knowing she's a traditionalist, she said, "When I go to the village, I see the number of people have doubled there. They are crawling all over like ants – but without food or money."

Our caregivers see a link between environmental degradation and public health. The mining industry leaves behind pools of standing water which causes mosquitoes to breed, so now we have an explosion of malaria in mining areas. In areas of deforestation, we have problems with yellow fever outbreak as people move between the forest and the degraded areas. In urban areas we have both medical and social problems relating to overcrowding, respiratory diseases, teenage pregnancy, HIV/AIDS, hypertension, diabetes, stress-related diseases and alcoholism. The slum areas have grown exponentially producing myriad street children who scramble to make ends meet and women get abused.

Ghana Health Service, [PO Box M44, Accra, Ghana, tel: (233-21) 666-151, fax: (233-21) 663-810, www.ghana.gov.gh/governing/ministeries/social/health.php].

Rofina Asuru

Rofina Asuru is the principal investigator for the Community Health and Family-planning Project (CHFP) which tests alternative strategies of delivering healthcare. The CHFP was launched to test the hypothesis that reorienting healthcare to community-based services could reduce fertility and childhood mortality despite local social, economic and ecological obstacles.

Initiated in 1994 as a three-village program of social research and strategic planning, the CHFP guidelines for a district-wide experiment were implemented in 1996. Existing community healthcare workers were retrained as paramedics and assigned to village residencies that were constructed and developed with community volunteer labor. With trained nurses in rural communities, clients have 24-hour access to basic healthcare and referral to specialized care off-site should the need arise.

The CHFP conducts community mobilization to gauge an individual community's reactions, collect data and disseminate health information. Ms. Asuru is one of three daughters and, as her parents never had a son, her father insisted she study hard in school so one of the children would be able to support their mother when he was gone. She obtained a degree in nursing and then later her MPH in Copenhagen.

Overcoming Misconceptions

When we began this project, we knew we wanted the nurse living in the community so we tested three living arrangements. One village gave an apartment in somebody's compound to the nurse. In another village the nurse happened to be the wife of the chief, so she stayed in her own compound. In the third village, they had a small a compound which somebody had donated and the nurse lived there. In the end, we learned a lot of lessons from these three arrangements.

The families housing the nurse in their compound complained that having the nurse live with them brought every sick person in the community to their house. The nurse married to the chief did not get clients, especially family-planning clients, because they were not comfortable going to the chief's compound. Having the nurse live in her own home in the community seemed the best arrangement, so communities offered to construct houses where the nurses could live.

Many research findings that have come out from Navrongo have influenced national policy. When it became apparent that imbibing large doses of vitamin A increased a child's chance of survival, that became part of our national immunization policy and now vitamin A is distributed alongside polio immunizations. We conducted a trial of netting over beds to see if this impacted mortality. It was clear that bed netting improved the chances of children surviving, so it has become part of our national policy. Finally we have shown that if healthcare is moved closer to the people, it increases child survival and it also reduces fertility.

I head the community health and family-planning project in seeking alternative strategies for delivering healthcare. In Ghana we have fixed facilities all over the country and we have nurses waiting there for clients to come for services. What we began to do was try and reorient this kind of service delivery. First we began using community resources like our chiefs, elders and leaders. This we had not done before. Everybody in the village was consulted and they were happy to give us their opinions about what we needed to do–they wanted to have the nurse in the village and were willing to provide her with housing and security.

The nurse, in exchange, offers them health services. The nurses are given motorbikes and are equipped with basic drugs in family planning. As they move from compound to compound, they visit various houses, depending on health needs, and will treat the community members for their minor ailments, or give them family-planning counseling or services. They discuss sanitation and other compound-related issues. Those problems they cannot manage, they refer to the next level.

This program has recorded some real successes. There has been a decrease in fertility by one child per family and an increase in child survival by 15%. The houses where the nurses live may not be the best and so we have learned that each year someone must go to help with maintenance. Basically maintenance is done by the women who are so saddled with household chores and their own maintenance, that by the time they can do anything for maintaining the nurse's house, it's the rainy season and they have to go to their farms. If we could provide more durable structures for the nurses' living quarters, that would help a lot, but many districts just do not have sufficient resources to build such houses.

The women in the communities are very pleased when we bring these services to them. From focus group discussions we've learned a lot from them. In the first place, the women are involved in all the work in the village–they go out to the farm and come home late in the evening–so many women come to the nurse in the evening to get their family-planning devices. With the nurse in the village, this facilitates such scheduling. We are trying to get the district assemblies to support the nurses' training, since they already generate their own revenue and get some from the government which is supposed to be used for developmental projects. They already support the training of teachers.

At the village level we have the chief or the chief in charge of the village, plus sub-chiefs or clan heads and compound heads. It's important when you are dealing with them to recognize their chain of command so you don't bypass anybody. We realized that men oppose the program when they don't have information. It's not that they want to be difficult, but when they feel left out of the loop, they get suspicious or hostile. In the past, men have opposed their wives' use of contraceptives because they worried women using contraceptives would not be able to have any more children or might become promiscuous. Much discussion with male groups is required to change such dominant mindsets in our society.

Navrongo Health Research Centre, Ghana Health Service, [Box 114, Navrongo, Upper East Region, Ghana, tel: (233 742) 22310, 22380, fax: (233) 21 22320, www.navrongo.com, e-mail: ahodgson@navrongo.mimcom.net, www.ghana-chps.org].

Dr. John Koku Awoonor-Williams, Ghana Health Service

After completing his medical degree, Dr. J. Koku Awoonor-Williams earned an M.S. degree in public health and returned to Ghana where he was posted in the Nkwanta district. He is the only doctor serving 187,000 people with a staff of two nurses. Brain drain is a major health problem for Ghana. Even with limited resources, Dr. Awoonor-Williams and his staff have increased child immunization to 73% and contraceptive prevalence from 2% to 16% for Nkwanta District. Dr. Awoonor-Williams is challenged on a daily basis to keep people alive. The community health program he established has quadrupled the number of women who have access to family planning and reproductive health services.

Reducing the Fertility Rate in Ghana

One of the challenges we face in Ghana is that there are women still having 10, even 12, children. When you ask them why, they tell you that they are not convinced that if they use family planning and have four children, at the end of their lives those four children will survive and they fear being left with no children. The tendency is for some mothers to feel that if they have 12 children, maybe they'll lose half of them but end up with six offspring. A major problem was to demonstrate to them that you can practice family planning and have your children live.

The Nkwanta district is very large with poor infrastructure development and no district hospital. I was posted there in 1996 as the first doctor ever to work in Nkwanta. I am still the only doctor working in this very deprived district. I can only presume that's why no other doctor wants to go there. There are no telephone facilities, the roads are almost impassable and even television is unavailable. If anybody's sick, the closest hospital to take that person to is seven hours away.

When I got there, family-planning prevalence was about 3%–which was an enormous challenge. In 1996 we started to provide health services in one of the nurses' living quarters. We had only two or three health centers so we need to find a structure that would provide a service point for those people coming for care.

Some 60 to 70% of my cases are emergencies. As the only doctor you have to do surgery, clinical work, see patients, handle public health issues and do administration. In 1996 we recorded an average of eight maternal deaths a month. Infant mortality was also extremely high. Any woman who went into labor who was unable to deliver the baby had to be transported 7 hours from

Nkwanta to the nearest facility. That in itself was the cause of much mortality for often those with obstructed labor simply could not survive the long trip to the hospital.

Several traditional beliefs compound the problem. One is that if you are pregnant and are faithful to your partner, then you should be able to stay in the room alone and deliver and should not have any assisted delivery. Consequently women will resist as long as possible getting any help so that by the time they bring a woman in labor to get help, it's normally very late. Transportation itself is primitive. Communities will improvise a chair which strong men will carry with the woman on the chair to the facility.

In Ghana we have tried a system of volunteerism, but it does not work too well. People volunteer for a year or so, but become fed up with the system. It is best when consensus is established between the community and volunteer to serve for two years. If they perform well, the committee reëlects them. This way the village health committee works together with the volunteer and the nurse. She treats minor ailments, provides immunization, health education, sanitation, home visits, family planning and does deliveries together with the traditional birth attendants.

The health committees now take the responsibility of providing maternal services. This program has raised our immunization coverage from 9% to 73%. Family-planning prevalence has increased from 2% to 16% which is almost equal to Navrongo. Nkwanta has become a center where district teams come for training. They are motivated by what they see. Last year alone we hosted about 13 district teams from Ghana.

Norplant insertion is a new method we introduced about two years ago and it is being well received in the community. People say they want to move from other methods to Norplant. At the moment, however, only doctors and the public health nurses are allowed by policy to provide Norplant.

James Phillips, The Population Council

James Phillips previously worked with the Council's international programs in the Philippines, Bangladesh and Thailand. His research focuses on measuring the demographic role of family-planning programs and conducting field experiments in reproductive change. Currently he is involved in collaborative research with the Navrongo Health Research Centre in Ghana in a program designed to assess the fertility and mortality impact of health and family-planning services in a rural traditional setting of northern Ghana. When a national program scaling up the Navrongo experiment was launched in 1999, Dr. Phillips served as an advisor to the Ghana Health Service, monitoring and evaluating a program named Community Based Health Planning and Services Initiative.

From Bangladesh to Ghana

Bangladesh was at the epicenter of a huge debate about the demographic significance of family-planning programs. It was argued that given the conditions of great poverty, high fertility, the low status of women, the absence of autonomy and so on, there was no prospect that a supply-side approach to service delivery could work. Nor could you offer family-planning services and expect a program to succeed. There had been a huge investment in the Pakistan era with all kinds of programs and strategies, but there was no evidence that any of this had lowered fertility levels. So the family-planning program was debated even though it was a program with huge external investment.

I went to work at the International Center for Diarrheal Disease Research in Bangladesh, a center for studying cholera. This was important, because the research machinery for studying vaccines could leverage research on family-planning's impact. You could go to an area where they had been working, an isolated, rural, impoverished part of Bangladesh, and know in advance the demographic characteristics of that population so you could track with great precision everything that was going on with fertility and mortality, migration, marriage and then use that resource to experiment with family planning to see what would work and what wouldn't. So that's what I did.

The notion there was no demand for family planning was simply false. I found there was a great, unmet need for services. But this demand was fragile. All kinds of offsetting, social circumstances prevented women from exercising any reproductive autonomy. If you got it right and organized a program that addressed not only the general demand for service but the specific needs for these to be provided in a particular way, then the impact was immediate and pronounced. Within two years the Matlab Model produced conclusive evidence that fertility would decline by two births of the total fertility rate.

The scaling up of the Matlab Model was supported by the World Bank and a key figure at the World Bank – Fred Sai – who was instrumental in building a consortium of donors for financing the large-scale implementation of the Matlab project so by the end of the 1980s this program was providing services to every doorstep in the country.

A Ghanaian medical scientist spearheaded the financing of this Bangladesh program. I met with him over lunch in 1991 and he asked me what it would take to replicate Matlab in Ghana, This led to an entire group of Ghanaians coming to Matlab and developing a research plan to replicate the Matlab experiment in Ghana.

Conducting Interviews with Ancestral Spirits in Africa

We looked at the literature on religion and all of the authors argued that traditional religion in Africa is a force aligned against family planning, that traditional religion reinforces pro-natalist ideas, that lineage is a sacred force with ancestors that cycle through the living world, pass into the after-world and govern daily lives. These ancestors are driven by the need to have as many descendants as possible.

In reviewing this literature, we found no empirical evidence to back up these claims. As a Western visitor to this area, I had the idea that we had to solve this problem by communicating with the traditional religious leaders, then educate these guys and get them on board for this program.

The Ghanaian anthropologist that I was working with said we had it all wrong. We shouldn't be interviewing religious leaders, but rather the ancestral spirits, since they are the ones who decide these things. So we went to lineage heads who are the spiritual leaders of extended families and interviewed them, using the reproductive preference section of the demographic and health survey.

The following day we went with them to a soothsayer and through the rites of soothsaying incantations and sacrifices – a lot of chickens were sacrificed for this study – we interviewed the ancestral spirits, asking the same questions we had asked the spiritual leaders the day before. The soothsayer then went through the rites and by some medium of communication would answer the question for the ancestor.

We then compared the two responses and discovered they were quite different, so we pointed out that the ancestors were not aligned against family planning and were not some anti-modern force, but were flexible and sensitive to social needs and far more consistent with women's responses than with men's. Through this we learned that religion is traditional but it's not a problem – it's more of a resource than a constraint. Western medicine and reproductive choices are not for the spirits to decide. The ancestors said this is not up to them. It was a matter of choice, so actually these spirits were pro-choice.

We learned many lessons from the Navrongo Project. First, family planning would work and the child health component of the experiment would work in a huge way. When you look at it more closely, though, there are some surprises. If you work strictly with volunteers, the program fails. There's an attitude in Africa in the donor community that you can build programs on traditional institutions without professional workers being involved in these programs.

What we found is that volunteers actually increased mortality. This was totally unexpected. The reason is that in a situation of great poverty a volunteer will divert parental health-seeking behavior away from clinical services or trained paramedics where they get antibiotic therapy for febrile illness, to informal procedures such as paracetamol treatment for febrile illness which fails to treat acute respiratory infection. So children die, even though they're getting care. This was a big shock because it was the cornerstone of UNICEF's regional policy for primary healthcare in Africa. This experimental study took on some conventional large-scale international initiatives and it proved quite conclusively they don't work.

Another surprise was showing that volunteers increased fertility. It turned out that contraception was often being adopted by women who were abstaining from sexual relations at the time. They were behaving not to control fertility but to change sexual relations from abstinence to active sexual relations with their husband. Since there is no immediate effect on fertility from the family planning method they used, it turned out the birth interval was shorter than it would have been had there been no family planning there in the first place.

The message was apparent: you have to provide services that have social acceptability plus a technology that promotes long-term use and avoids a risk of embarrassment to the women who adopt it. You also have to do a lot of work with men to build their partnership and a sense of commitment to the program. The key that made the Navrongo experiment work was the social mobilization of men.

In the matter of female circumcision, or female genital mutilation (FGM), marriage and family building has a system of putting together wealth in the form of animals to give to another lineage to secure the marriage of a young woman who will then come back to the parent lineage and produce children. The couple is incidental to such social arrangements which lead to polygamy and decision-making that doesn't involve the young women. But women are then put into a multi-extended family unit called a compound. Often there are multiple wives in the marriage. Sexual rivalry in this context is considered fatal to the tranquility of the extended family, so women must subdue the sexuality of other women in the family. They believe that female genital mutilation achieves this and that an uncircumcised young woman coming into an extended family where everyone else has been so treated represents a threat to family tranquility, stability and values.

Women also believe that this is a rite to womanhood. One becomes a woman through the rites of circumcision, so a woman who is not circumcised is crass, not cultured, does not have the right values, might be promiscuous, might be a sexual misfit and can't be trusted–and so on. Peace, tranquility and upstanding womanhood depend upon FGM. Finally, if a woman dies, her eldest daughter performs the rites of the funeral. If the eldest daughter is not circumcised, she's viewed as manly and not a proper woman and unfit to perform the rites of a funeral. That's a terrible disgrace for a woman to be denied. An uncircumcised daughter can't even attend the funeral, let alone practice this rite and no woman wants to face this consequence.

Listen. Sit under the Banyan tree with men, women, elderly women and chiefs and listen. Explore what people are willing to support, what their concerns are and what the way forward might be. In this case we found that men are the agents of social change of female circumcision. They don't have a heavy investment in the practice. They're powerful in society and they're powerful in the family. They're capable of organizing social events and community educational encounters, yet they have no emotional attachment to the practice. So you build the program around the leadership of men and then legitimize change through activities with women. Women have networks, lending networks, singing groups and various social groups that they form for economic, social and health reasons, and most important are these lending societies.

We formed social-change groups to train women in matters that would help them with microcredit and other activities that they really wanted. Then we built the theme of the dangers and risks of female genital mutilation into this program. You don't just start an educational campaign. Figure out what people want and then integrate a program to achieve social change in the context of this other more general programs that they are supporting and you're working with them to develop.

In Ghana hundreds of firms go to large-scale organizations and change the way things work. Often, while they are effective in what they do, their techniques and methods don't necessarily translate to the health and population field. What works in the family-planning field is not so much what organizations do, but what couples do outside the boundaries of an organization. The organization, to be effective, has to be fundamentally adaptive to cultural realities and organized so it looks like and sounds like the society where it's based.

You must listen to people talk about what they want. The first principle is that children are dying and the parents are intensely concerned about child health. To address the society's concerns about fertility without addressing this antecedent issue is an invitation for disaster. I thought it would be interesting to sit with a group of women and then chiefs and elders and men talking about the ICPD, the International Conference on Population Development in Cairo–the themes of that conference, and what was the view from the Banyan tree on this broadening of family-planning programs that was being proposed with gender issues as the forefront of change.

The uniform opinion was that this was incredibly narrow. The ICPD could never work because it wasn't broad enough. What people needed and wanted was basic healthcare and basic survival – plus some assurance that if there was a crop failure or some disaster, they were going to survive. Then you could talk about these other things. That was quite a shock to me, because I thought we were broadening the perspective from the point of view of people under the Banyan tree, yet we were only taking a half-step.

For us, a big problem is that donors are very impatient. They attempt to do something new, something bold, but do not recognize that maybe these programs are going to take a generation to work. It takes investment plus wisdom to support a program that can't be done in five years.

For over three years, I lived in Nigeria – one of the great political tragedies in Africa. With 450 ethno-linguistic groups and three huge ethno-linguistic political blocks, this country has profound forces pulling it apart so it's difficult to develop a centralized program. Bureaucracies lack historic grounding. Political parties lack integrity. All tertiary institutions available in Asia to make such big programs work are not in place in Nigeria.

The strategy we developed in Ghana might indicate what would work in Nigeria is developing a program creatively based on decentralization and diffusion of innovation at the periphery. Finding ways of channeling resources directly to localities and avoiding these big central bureaucracies will never work. It's a political task that's going to be a difficult one to resolve, but I think Nigeria is a case study in the political dilemma of how to succeed with these programs in Africa. Ghana, on a small scale, represents a case study in how to surmount those difficulties and develop a program that works.

Population Council, [One Dag Hammarskjold Plaza, New York, NY 10017 USA, tel: (212) 339-0500, fax: (212) 755-6052, e-mail: pubinfo@popcouncil.org, www.popcouncil.org].

Marcela Villarreal, **Food and Agricultural Organization, United Nations**

In 1988 the Food and Agricultural Organization of the U.N. (FAO) began analyzing the effect of the AIDS epidemic on rural development and food security. The generation hardest hit by the epidemic is also the most productive in terms of agricultural labor. So far an estimated 7 million agricultural laborers have died due to AIDS, and another 60 million could die within the next 20 years. The deaths of these workers undermine the family unit in many ways. Without the principal breadwinner alive, food security is endangered for the older and younger generations. Many children are orphaned or rely on their aging grandparents for support.

Quite often this leads to child prostitution, which is a function more of basic survival. The disease is spread not only through survival prostitution, but through the intransigence of the agricultural workers who travel from location to location, and often through the practice of leveret. Leveret is an African custom that forces a widow to marry into her deceased husband's family in order to retain her husband's land. If she refuses to do so, she will not be able to grow food.

FAO is currently working on an agriculture sector strategy to mitigate the impact of HIV/AIDS. Initiatives include labor saving technologies within the field such as zero-tillage techniques, energy-saving stoves that use less time to fetch wood, which primarily is a job for females, or irrigation projects that bring water to households and fields. Women are the chief caregivers in the African household; taking care of a family member with AIDS drains time from the daily chores that need to be maintained. Marcela Villarreal's emphasis at FAO has been on HIV/AIDS programs.

HIV in Africa

Today we are seeing devastation and in the rural areas the epidemic is impacting both livelihoods and food security. There when you ask people what they need, the first thing they say is,

"We are hungry." More important to them than medicine or other help, is receiving food aid. HIV/AIDS has impacted food security in rural areas because it impoverishes people. Survivors divert household resources to take care of sick relations – and then pay for funeral expenses. They have less labor power within the household because HIV/AIDS is killing off those in their most productive years, so rural households are not able to produce enough food.

Survival Sex

An unfortunate situation growing in rural Africa is that people who are hungry have no other resource but to sell their bodies to be able to eat – one day at a time. Providing food security is thus a way of preventing the further spread of the epidemic through what we call survival sex rather than prostitution. That's why the FAO is working to develop labor-saving technologies for the agriculture sector to mitigate the impact of HIV/AIDS and prevent the disease from spreading. The agriculture sector desperately needs labor-saving strategies not only in the fields, but also in homes.

We're also working on the preservation and transmission of agricultural knowledge. In many African communities parents are dying faster than they can transmit agricultural knowledge and skills to the huge number of orphans they are leaving behind. We must help this orphaned generation figure out what can be done with less labor that will provide enough nutrition for health plus giving proper nutrition to those living with AIDS. We are promoting home gardens and agricultural diversification that needs less input of work, fertilizers and pest controls, but can still provide nutrition and a variety of foods for the needs of the household level. There's a whole generation that's being wiped out because of the epidemic, leaving behind only the elderly and the orphans.

A Personal Mission Statement

One experience really touched me many years ago when I was

just starting to work in rural Honduras, an extremely poor region in a poor country. I interviewed a woman, pregnant with her eighth or ninth baby, who was crying. She was with a friend who asked her, "Why don't you use family planning?" In tears, she replied, "I couldn't. My husband would kill me." Suddenly I realized this woman was my age. I was fresh out of graduate school with my Ph.D., hoping to do something to help people and I realized that my life and hers were completely determined by different realities and that in order for me to do anything that would be meaningful to her, I would have to make a real effort to understand her own point of view and her own priorities.

Another time in Delhi, India, my taxi driver let me off at the wrong place. I walked 300 meters and in that short span I saw poverty of a kind I couldn't imagine was really possible. Feeling it was so unjust human beings should have to live in those conditions, I sat down and began to cry. These experiences changed my outlook on the world.

Food and Agricultural Organization of the United Nations (FAO), [Viale delle Terme di Caracalla, I-00100 Rome, Italy, tel: (39 06) 5705-2346].

6

Europe

Introduction

The three European countries visited during the filming were France, the Netherlands and Italy. Like most of Europe, their fertility rates are less than replacement, but only Italy is projected to slightly decline over the next 50 years from 58 million to 52 million. During this same period the Netherlands is projected to add one million to their 16 million and France four million to their population of 60 million. These increases will partly be due to immigration, which was discussed briefly in the interviews. In the Netherlands the portion of the population over 65 will increase from 14% to 24%. All of Europe is concerned with the impact on the environment of an ever-increasing, prosperous population and the potential threat of interrupted flows of fossil fuels. For this reason they are focused on improving energy efficiency and finding alternative energy supplies, such as wind or solar.

Many European countries have increased their support for the United Nations Fund for Population Activities with the Netherlands being the largest contributor. Governments and the private sector have been successful in preventing teenage pregnancies and providing access to family-planning/reproductive health services. The declining population causes less concern than how immigration will factor into future labor needs.

Dr. J. J. van Dam, Rutgers NISSO Group

The Rutgers NISSO Group (RNG) is a merger between the Dutch Family Planning Association and the Netherlands Institute of Social Sexological Research (NISSO). RNG is a center focused on sexual reproductive health and its mandate is to achieve a society where individuals have the ability and opportunity to make voluntary, well-informed choices with respect to their sexual conduct and personal relationships. It also disseminates information geared toward prevention of sexually transmitted diseases (STDs), unwanted pregnancies, sexual discrimination and sexual abuse.

Dr. J. J. van Dam is a medical doctor and head of the department of innovation and development at RNG. The Netherlands has a progressive view of sexuality, yet Dr. van Dam is concerned about the increase of unwanted pregnancies, STDs and HIV/AIDS, although AIDS is not a major concern yet in Holland. Now 14% of Holland's population is over 65; soon 24% will be over 65. The aging of population is a major concern for all Europeans.

Every Child a Wanted Child

I wish every child were a wanted child. When I started my career, I wanted to be a gynecologist and worked in a hospital where I performed many deliveries. I also worked in an abortion clinic, and that increased my interest in providing information about the planning of one's family, about contraception and the prevention of unwanted pregnancies.

In the 1970s and 1980s, you saw many one-parent families because couples got divorced easily. I think now there is an adverse reaction to this. People want to get married and want to have a stable family life again and also more children.

We have a tradition of talking very openly about sexuality here in Holland. Sexual education is given at schools and parents talk openly about sexuality with their children. The main theme in discussing sexuality here in Holland is safety. Some people feel this is not the right approach anymore because with so much emphasis on protecting yourself against STDs and unwanted pregnancies, we forget that sex also has to do with love. Maybe we need to begin looking again for an approach to sexuality that is fun and that reawakens the passion.

Rutgers Nisso Group, [Postbus 9022, 3506 GA, Utretch, The Netherlands, tel: (31 30) 231-3431, fax: (31 30) 231-9387, e-mail: rng@rng.nl, www.rng.nl].

Dr. Wouter Meijer, the World Population Foundation

The World Population Foundation (WPF) supports local organizations in developing countries which design and manage sexual and reproductive health projects. They also provide technical advice and training. An area of concern to WPF is to promote groups that prevent female genital mutilation. WPF has expertise in reproductive rights and reproductive health, delivering quality reproductive health and family-planning services, information, education and communication on sexuality and reproductive health programs for young people, and strategies to achieve sustainable development. WPF distributes information to the media, NGOs and the general public and organizes seminars and training programs in the Netherlands and other European countries.

Dr. Wouter Meijer was the executive director of the WPF from 1992 until 2003.

The North/South Population Divide

The World Population Foundation is a Netherlands NGO that works to promote reproductive health and rights worldwide, exclusively in and for developing countries. We promote these rights for everyone, but we concentrate on young people – girls and boys – so that those in developing countries can realize and implement the same rights that we find totally acceptable and normal and don't even think about any longer in our rich northern countries.

A key issue we confront is the unmet need of about 300 million people in the developing countries who have no access to appropriate services, information or counseling nor to quality care in this area of reproductive health. Another striking issue is maternal mortality, one of the most shocking indicators of the divide between the poor and the rich countries – 99% of this takes place in the south and maternal mortality at present stands at 1,500 women dying in pregnancy or childbirth-related causes every day, or about 600,000 every year.

Thinking about Numbers

By looking at the numbers and talking about population growth, one can observe how thinking and policies on numbers have changed dramatically during the last two decades. In the late 1960s we saw the Club of Rome and the doomsday thinkers who thought population growth was unstoppable, that world population would never stabilize and towards the end of the 21st century this would produce some sort of doomsday, with maybe 30 or 40 billion people in the world. Of course, there is no way this fragile planet could ever offer the resources for such masses.

But dramatic changes have altered these doomsday scenarios. Year after year, it has been demonstrated that if people, especially women, can freely choose and have access to the means to implement their choice, they opt for lower fertility. Nonetheless, we're likely to still add almost three billion more people to the planet, which is equivalent to the entire world population of 1960. This is why we should meet the unmet needs as soon as we can, in as many countries and on as large a scale as we can.

The Pledges at Cairo

What is good for people is good for the world. In 1994 at the ICPD in Cairo, the world united on an unanticipated successful consensus and 180 countries signed this document. The program of action at Cairo brought a watershed change from policies driven by demographic targets to those which were based on human rights and oriented towards meeting the real needs of people: reproductive freedom of choice. That was dramatic. Many countries have followed suit, both in the donor community and certainly in the developing countries. But these improvements are not enough.

We need to fully meet the commitments that nations agreed upon at that 1994 conference (the 20-year plan of action). So many beautiful final documents are created at these conferences, but we don't meet our pledges and it's very sad and embarrassing to observe that the donor countries that were so enthusiastic about Cairo today are fulfilling only about a third of their commitments. They had promised 5.7 billion dollars a year by 2000. They're barely coming up with two billion.

The developing countries, on the other hand, are actually doing better and are meeting some two-thirds of their commitments. Overall, less than half of the Cairo promises have been met. While we can observe that Cairo's objectives work, they are being implemented far too slowly. Thus our advocacy task is so important. We need to urge everyone to stop talking about the problems and the policies–all well-known factors–and put our money where our mouth is and act. This is the almost verbatim quote of our last development minister, and we in the NGO community cannot agree more. In the 1990s the world defined the key global issues with little disagreement. Now it's time to act and stop gathering at these massive United Nations conferences.

Love

I don't see how you can educate young people about the mechanics of sex without addressing the link to love. That's really where you hit them–when they are 13, 14 or 15 years old–both boys and girls. When you talk about that link between love and sex, you cannot leave out love.

World Population Foundation, [Ampèrestraat 10, 1221 GJ Hilversum, The Netherlands, tel: (31 35) 642-2304, fax: (31 35) 642-1462, e-mail: office@wpf.org, www.wpf.org].

Dr. Gijs Beets, **Netherlands Interdisciplinary Demographic Institute**

The Netherlands Interdisciplinary Demographic Institute (NIDI) is a research institute of the Royal Netherlands Academy of Arts and Sciences (KNAW) engaged in the scientific study of population. Research carried out at NIDI aims to contribute to the description, analysis, explanation and prediction of demographic trends in the past, present and future. The determinants and consequences of these trends for society at large and policy in particular are also studied. NIDI research is characterized by its interdisciplinary approach and international orientation. Gijs Beets, a researcher at NIDI, is an expert in social geography and fertility.

Contemplating Dutch Demographics

The population in the Netherlands is over 16 million. We debated in the 1960s how we would accommodate so many people and asked whether there should be some discussion, or even policies formulated, trying to prevent so many people from living in the Netherlands in the 21st century. The final report of the state committee was not so much about how to accommodate over 20 million people, but how to deal with a significantly aging population within the borders of the Netherlands which today is slightly less than the average in the European Union, but it is growing to about 24% in 50 years' time.

It doesn't seem many people are thinking about the implications of this aging demographic or realizing what it will mean for countries like ours–even though this is an international issue because the same trends are being seen in other countries. No country, as far as I know, is not experiencing an aging trend. Even the countries in Africa are aging. The consequence is that when you walk on the street you see few, if any, children but many older people. This is a situation that we have never had before.

The consequences this will have on health insurance premiums, let alone political power, are difficult to predict. Of course, we have had the Roman Empire, which came and went, and we had Europe in the Golden Age (17th century) when Europe was the world leader. Now the U.S. is powerful. Here in the Netherlands there is no population policy. As a demographer, I would hope that in the coming decades we could stabilize demographic trends as soon as possible. I really think that a world with two billion inhabitants is much better off than one with six billion, and certainly better than nine billion. I am not even thinking about the possibility of over ten billion inhabitants in the world. That would be a challenge.

Gijs Beets, NIDI, [PO Box 11650, 2502 AR The Hague, Netherlands, tel: (31 70) 356-5200].

Evert Ketting, Netherlands School of Public & Occupational Health

The mission of the Netherlands School of Public & Occupational Health (NSPOH) is to improve the health of the population. NSPOH provides a high level of professional and academic training and supports public health organizations by integrating research and educational activities performed in different professional and academic settings. Evert Ketting is a staff sociologist at NSPOH.

The Dutch World of Family Planning

Until 1970 the Netherlands was the most densely populated country on earth. Then, it was bypassed by rapidly growing countries like Bangladesh and Taiwan. In the 1960s we had one of the highest population growth rates in Europe. Contraceptive education improved, and as a result the birthrate went down more rapidly than it ever had in any country during such a short period.

Around 1970 contraception became available free of charge in the Netherlands. In 1970 our parliament decided that medical contraception – the pill, an IUD, sterilization – should be included in the national health insurance scheme. The government then actively stimulated public education. Service delivery in the area of contraception in the country became available through the family doctor, that is, it was available just around the corner. You didn't have to go to a far away family-planning clinic but just to your own family doctor. All Dutch citizens have their own family doctors and know them very well. This doctor is very accessible and there is a high level of trust in the family doctor, so the combination made the transition to effective contraception easy.

The Netherlands was one of the initiators of the United Nation's Population Fund in the late 1960s and has always been extremely supportive of UNFPA, both in terms of technical assistance and financial contributions. In fact, it is now the largest single contributing country to UNFPA, although it is a tiny country compared to, for example, the U.S., Germany or Japan. In terms of per capita, the Netherlands is very close to Norway, which is still the per capita first contributor to family planning.

The general perception of the Netherlands is that our country is almost radically tolerant, and sometimes anarchistic. That's half of the story – and more of an outside perception. The inside story is that the Dutch are very determined about regulating their behavior. Looking at issues like sexual health, contraception or family planning, Holland is very successful. We are a model country because we have the lowest abortion rate in the world, we have the lowest teenage pregnancy rate in the world – all these indicators are extremely positive in the Netherlands. This is largely explained by the open atmosphere regarding issues of sexuality.

While the success of the Netherlands cannot immediately be replicated by other countries, some lessons can certainly be gleaned. First, what we have seen in countries like India is that a top-down approach does not work. As soon as the government tries to impose fertility regulation targets on the population, its policies become counterproductive. The lesson we learned in the Netherlands is that an effective approach has to be bottom up–trying to identify the needs of the people and then responding to these needs effectively while involving the people in that process. Second, it is important these services are delivered close to those who need them. Both psychologically and geographically, and of course financially, the barriers must be low. This is one of the reasons behind the success story of the Netherlands.

The rapid decline in the average number of children per family started to happen in the late 1960s when universal education was available and when women's educational levels were approaching those of the men in our country. This has definitely been a factor, because if women are not educated, they will never be able to speak on behalf of their own interests. So a key factor is that education is absolutely crucial. Those countries with low educational levels where women are not treated unequally are countries where population growth is still rapid.

North/South

I think the global solidarity between rich and poor countries is diminishing. This is a very serious problem. The commitment of rich countries to invest in the development of poor countries is comparatively low and as a consequence, the gap between the rich and the poor countries is growing wider and wider. It's not only for them, but it's also for ourselves that we should invest much more. I'm not talking about 2% or even 5%, but rather 200% more in helping with the development of the poorest nations. It's the only way to solve these kinds of problems.

Immigration

We in Western countries are much worried nowadays about illegal immigration. It's a key political issue in all Western European countries. Some feel, but I don't agree, that they are being flooded by so-called refugees from developing countries and that they are losing their own identity. In the Netherlands, this anti-immigration sentiment has increased rapidly these past few years, but we have learned that simply tightening immigration regulations will, in the end, not work. We live in an open Europe and increasingly we live in an open world which simply cannot be closed off. The only real answer is to reduce the pressures that drive people from developing countries to rich countries. In the poor countries we must help foster sustainable development in those countries themselves. It is the role of rich countries to invest in development abroad to diminish the kinds of problems many people worry about.

Holland has been recognized as a success story, but we have failed in bringing our new immigrants up to the same level as the original Dutch. If you look at all the indicators–sexually transmitted diseases, unwanted pregnancy and abortion rates–the immigrants are worse off than others in the population. In fact, the recent increase in the abortion rate is chiefly the result of new immigration to the Netherlands. That's an issue we haven't yet dealt with properly.

The Environment

Environmental issues are linked to the population growth issue, but mainly these are linked to our consumption patterns. We are spoiling our own environment, not because we have too many children here in the Netherlands, but because we produce too much waste. On the other hand, the Netherlands is a densely populated country and our food production is enormous. We are one of the largest exporters of vegetables, fruits and dairy products in Europe. So we are not dependent on food from abroad. So that is not the problem.

Evert Ketting, Ph.D., [Regentesselaan, 3708 BM Zeist, Netherlands, tel: (31 30) 692-3083].

7

United States

Elizabeth Laura Lule, World Bank

The mission of the World Bank is to fight poverty and improve the living standards of people in the developing world. It is a development bank which provides loans, policy advice, technical assistance and knowledge-sharing services to low- and middle-income countries to reduce poverty. It also promotes growth to create jobs and to empower poor people to take advantage of these opportunities.

The World Bank works to bridge the economic divide between rich and poor countries. As one of the world's largest sources of development assistance, it supports the efforts of developing countries to build schools and health centers, provide water and electricity, fight disease and protect the environment. As one of the United Nations' specialized agencies, it has 184 member countries that are jointly responsible for how the institution is financed and how its money is spent. There are 10,000 development professionals from nearly every country in the world who work in its Washington DC headquarters and in its 109 country offices.

The World Bank is the world's largest long-term financier of HIV/AIDS programs and its current commitments for HIV/AIDS amount to more than $1.3 billion–half of which is targeted for sub-Saharan Africa.

Elizabeth Lule, a Ugandan, is an advisor to the World Bank who decided to dedicate her life to work for the improvement of women's rights when she was teaching women to read and write. This basic education exposed women to information that helped them make important choices in their lives and gave them empowerment.

World Bank Population Initiatives

As an adviser to the World Bank, I am responsible for working with countries to address the impoverishing effects of population growth. We are engaged in policy dialogue with countries, conducting analytical work, building capacity within the bank and also in each country, to recognize population and reproductive health issues in the development agenda.

One effort in India funded by the bank is the reproductive and child health project, discussed above, which recognizes the connection between maternal and child health. Healthy mothers make for healthy children and healthy nations. The bank is increasing access, building capacity at the state levels, but also working with NGOs to make sure the communities are involved. Improving education and the status of women is vital to their making choices about the number of children that they want to have. In the future the bank plans to focus on increasing male participation and involvement. Women cannot make the choices by themselves.

The age at marriage in India is still low. Young girls marry and bear children very early, too early for their own health. By focusing on birth spacing, we help improve the health of young mothers and their children. I think keeping girls in school is critical because this has shown a demonstrable link to a later marriage. Women can make better decisions and reach for personal rights in a way young girls cannot. Education also improves their motivation and knowledge about medical assistance for themselves and their children.

The bank recognizes the importance of NGOs as entry points into communities. Communities are really the agents of change, especially when we're talking about reproductive health issues, because many of those decisions are ingrained in the norms and values of those communities. People cannot change their behaviors unless they have a social enabling environment. We fund some NGOs directly as a way of expeditiously addressing some of the difficult issues like female genital mutilation or sensitive adolescent concerns and contraceptive technologies.

Case Study: Africa

Access remains a big issue, because women in rural areas have poorer access to healthcare and family planning. It may take a whole day to walk to the nearest clinic and once you get there, you may find no drugs, supplies or contraceptives available because the health systems in many African countries remain weak and fragile. Human resource issues and lack of providers are still very important in Africa, a region currently in the throes of a debilitating brain drain in relation to its medical providers. Doctors and nurses are moving north where they can earn better salaries and have a better quality of life in many northern countries experiencing aging populations. This is a critical issue for Africa. The status of women in Africa also is a major problem. Low literacy rates, lack of knowledge and information about family planning and even protecting themselves from HIV/AIDS is a big challenge.

My country of Uganda is a success story. The political commitment, the openness of discussing HIV/AIDS and focusing on vulnerable groups like young people, plus the involvement of communities in addressing this issue have come together to produce this success. Faith-based groups or the stakeholders have taken it upon themselves to address these matters, so infection rates have declined among young people who have reduced their number of partners. The age at sexual debut has increased. Of course, there are capacity issues that remain difficult for Uganda, but at least the leadership has in place an appropriate policy to face these problems.

HIV/AIDS is an emergency situation for Africa and most of the bank's lending is for HIV projects. Nevertheless, HIV is a reproductive health issue so a lot of the work we are doing now is trying to strengthen the linkages between HIV/AIDS and the need for family planning as well as other reproductive health issues.

The Power of Sacrifice

Growing up in Uganda, I witnessed the heavy burden women had from the number of children they had plus the high infant mortalities they suffered. As I progressed through school, I was always motivated to help other women to ensure they had the information and the knowledge necessary to make informed choices. What convinced me that this was an area worthy of dedicating my life to was when I was teaching women how to read and write, I was amazed how this could empower them by exposing them to the information they needed to make the most important choices of their lives.

The World Bank, [1818 H Street, N.W., Washington, DC 20433
tel: (202) 473-1000, fax: (202) 477-6391, www.worldbank.org].

Thoraya Ahmed Obaid, United Nations Population Fund

The United Nations Population Fund (UNFPA) is the leading international agency working in a number of reproductive health areas. UNFPA focuses on safe delivery, family planning, prevention of HIV/AIDS, female genital mutilation and the prevention of violence against women. Recent emphasis has been directed at adolescent reproductive health issues. The current adolescent generation is the largest in the history of the planet and many in this target group have no access to information affecting their reproductive and general health. To deliver its services, UNFPA collaborates with national governments trying to respect indigenous customs and traditions.

Thoraya Ahmed Obaid, executive director of UNFPA, says the mission closest to her heart is the eradication of global poverty by providing for the basic needs of the poor, including education, health, reproductive health and gainful employment. From 1998 to 2001, Ms. Obaid was director of UNFPA's division for Arab States and Europe. Before that she worked for the Economic and Social Commission for Western Asia (ESCWA) in the social development and population division and was an affairs officer responsible for the advancement of women.

A Wide-ranging Program

UNFPA has 142 programs in Latin America, the Caribbean, Asia, the Pacific, Africa, Arab States, Eastern Europe and Central Asia. World governments have agreed that by now 90% of the 15-24 year-olds should have access to both information and services related to reproductive health, so we work with governments to set up such programs, which would include supporting those government's effort to train health workers and learn how to deal with young people – which is a sensitive area. Learning how to be effective in counseling them, providing them health services that include reproductive health and responding to their needs so they can protect themselves from sexually transmitted diseases is essential. Also included is to provide counseling and testing, as well as condoms in cases of protection from HIV/AIDS.

Historically UNFPA has been sensitive to the communities where it works and coöperates with our national partners, governments, NGOs and international partners to implement programs in the communities. Therefore, we place special emphasis on the culture or context in which the programs are being set in motion, so the messages that are of an international nature – like human rights and the rights of young people to access information and services – are transmitted through local languages and through the understanding of the community. We are tying the needs of young people and their rights, within a universal rights concept, into the communities and the local setting.

It's important to note that there are some 7,000 young people who are infected every day with HIV/AIDS – the majority of whom have no information about it and don't know how to protect themselves. UNFPA is the only organization working with adolescents in a comprehensive way of dealing with HIV/AIDS and other sexually transmitted diseases in the communities. Adolescents face many risks: teenage pregnancy, unsafe abortions, gender violence and HIV/AIDS remain serious threats for young people in the developing countries.

UNFPA addresses these challenges by working with governments and NGOs to establish programs and raise awareness among both young people and adults. We respect cultures and work with all partners to ensure that human rights are upheld and understood within that cultural context. UNFPA is part of the international community that at the Cairo Conference on International Population and Development, adopted what has been called the Millennium Development Goals: to reduce poverty, maternal mortality, HIV/AIDS, empower women and ensure adolescents are protected and have information and services in the area of reproductive health.

Today's burgeoning youth population is a powerhouse which if provided with education, good health and employment opportunities will be able to take us into the future. If we do not address their needs, we are only creating a generation that has aspirations but who have been stunted. A frustrated people, combined with poverty, unemployment, ill health and a lack of education creates a disaster. We have a moral obligation to ensure that we fulfill their human rights as citizens of this world, regardless of what their age is.

Environment, Population and Women

The environment and population are in direct relationship. Our hope is to have enough resources in our natural environment to be able to support the human population with a quality of life where basic needs for shelter, food and water are all satisfied. Within that perspective, women have a special place. In Africa they are responsible for fuel gathering and the ones who must see that the needs of the household are met. Women are often responsible for bringing water into the house and are pivotal in many areas of the environment – like keeping a house in rural areas free of infection and disease.

But women's needs are not being fully met in the world. We still have high maternal mortality rate and many women who are ill because of being pregnant. Women should not have to face death because they give birth. By providing women with education, we bring light to their lives, empowering them to make decisions about their immediate life, their family's situation as well as the community at large. Education also enhances their perspective on their life and helps them feel as though they are not victims anymore, but rather citizens and participants. This gift to women is essential to any development effort.

The Cairo Conference

The commitments made by governments in Cairo to provide significant funding for population issues, specified that two-thirds would be provided by the program countries themselves and one-third would be by the donors. So far the developing countries have come up with most of the funding – about 70% – while the donor countries have reneged on over half of their commitments. Many excuses have been given regarding the economic crisis in different places of the world, though to be fair, some developed countries like Denmark, Norway and Sweden have increased their contributions. The Netherlands has done a wonderful job of providing resources and so has Japan. Unfortunately, what's at stake when funds are cut or stifled, is not whether UNFPA will exist or not, but whether women will die or not.

With globalization, what happens in one part of the world affects the rest of the globe. Where there is poverty, there is unrest. Young people are lost and have no idea what they can do with their lives, no true prospects when there's no employment. At the same time, they see wealth all around them and the contradiction between the North and South becomes a great digital divide. There is also a food divide, a natural resources divide, a health and education divide–all of which impact not only on the poor, but the wealthy populations in developed countries, one way or another.

Needed Action

Population is not just about numbers. It is about people–and the quality of life. Therefore if we don't act quickly there will be more children than the world can accommodate or the environment can sustain and there will be suffering. That's why we must balance the human population with the environment.

United Nations Population Fund (UNFPA), [220 East 52nd Street, New York, NY 10017, tel: (212) 850-5601, fax: (212) 370-0201, www.unfpa.gov].

Joseph Chamie, United Nations Population Division

Established in 1946, the United Nations Division of Population is part of the department of economic social affairs. UNDP addresses population as it affects the broad agenda of secretariat issues relevant to the fertility, mortality, migration, projections and population policies. Information on these issues is disseminated to governments, the public, and scholars. Publications issued by the UNDP include *Population, Environment and Development* which lists each country and various environmental indicators as they relate to water consumption, motor vehicles, poverty, undernourishment, cropland area, forest area and problems associated with urbanization and international migration. Joseph Chamie is director of UNDP and his long tenure in the health and family-planning sector began when he and his wife joined the Peace Corps in 1967. Stationed in Bihar, one of the poorest states of India, Mr. Chamie became committed to population issues.

Principles of Population

During the 20th Century the world's population grew from about 1.6 billion to about 6.1 billion–a great deal of this expansion occurred in the developing world, wreaking a huge impact on the environment. Opinion surveys show many people are most concerned about the environment–and prepared to pay more to keep the environment sound and to clean it up. But the measures taken must not dramatically affect the economic situation in countries because then you start affecting jobs and the welfare of households. *That* is politically difficult to pull off. People want to work, yet at the same time they want to have clean air, safe water and the resources for their grandchildren's survival.

In 98% of all countries, women may legally have abortions if their life is threatened. Generally most countries have concluded they wish to minimize the need for abortion. The controversy hinges upon how to do that. One way is to make family planning more readily available, permitting couples to choose the number and spacing of their kids. Then they may not have to resort to abortion, and I think that's what the goal is of all countries–to reduce the need for abortion.

Demographic predictions

In the early 1950s when demographers were projecting what would happen with the world's population, they anticipated the decline of mortality causing a large difference between the birthrate and deathrate which led to some very explosive population growth in many countries. Since then birthrates have come down and mortality has decreased, but with the unpredicted issue of AIDS we see a leveling off of the growth rates to about 1.2% today. At its peak in the 1960s, it was 2.1%, so we have a decline of the gross rate. No one predicts a doubling of the world's population in the near future, but rather current estimates see a population stabilizing for the world as a whole to somewhere around ten billion by the end of this century.

This is a decline from recent estimates as high as 11 or 12 billion. But declines in fertility in countries like Iran, Mexico, Brazil, Malaysia, Thailand, Turkey, Morocco and Tunisia occurred much more rapidly than was ever expected. No one can predict with certainty because people are unpredictable. If people were able to achieve the number of children they want, this high estimate could be reduced quickly.

That's where family planning comes in–and why at the 1994 Cairo conference it was agreed that a basic principle was to permit couples and individuals to have the right to choose the number and spacing of their children and also to have the means and information to do so. Were that to happen all over the world, everyone would probably be wrong in their projections. As a scientist, I'm one of the few that always hopes my projections are wrong.

The world must adjust to its increasing numbers in a way that's sustainable and does not overly deplete resources and does limit damage to the environment. Clearly we have made mistakes in the past. Everyone has to be responsible and contribute to making the environment better. Governments have a special responsibility because elected officials and their policies, taxes, regulations and ideas are the ones that actually drive our behavior.

The Family-planning Solution

When people come to family-planning clinics in India, Brazil, Nigeria or Indonesia and are provided with these methods, they don't need to come asking for an abortion. Of family-planning methods there are many options to choose from, depending on what is appropriate for health, culture or religion. Our goal is to develop a world where every child is wanted.

Population has many new problems. For example, sex ratio: should you be able to choose the sex of your child? Now in many areas of China and India, because of sonograms, people are choosing selectively, aborting fetuses so they can have more sons than daughters. That's going to create a demographic imbalance. If there aren't sufficient brides for the young men, there's tension. Some European countries are facing shrinking populations – they're going to be smaller and older – as is Japan.

There is no single population problem. Of course a larger population is a major concern as Thomas Malthus predicted, but there are many dimensions of this. Besides the health issue, there is mortality and the AIDS crisis. New diseases, viruses and microbes could easily impact the world's population. Then how do we deal with aging? Life expectancy now is relatively low compared to what people are talking about. Should people begin living into their 100s, how will societies deal with this? You retire now in your 60s, but you may have to work to a decade or so longer.

In a few years, we will be reaching a new population milestone where for the first time more than half of the world will be living in cities. With the majority of the population as urban dwellers and living in crowded areas, how will we cope with this? Now we have cities as large as some countries with 20 or 30 million people. How will these cities get the water, the food and the infrastructure they need? How will we deal with them politically?

In some societies such as Germany, Finland and the Netherlands, 20% of the women in their 40s are childless. They have opted not to have children and there are opportunities now for women to have a great deal of say about their lives in terms of what their role will be. What kinds of opportunities or life-styles are they opting to have? The changing status and role of women is another transformation permeating every corner of the world. You also have differences occurring in societies because of new waves of immigrants. Food changes. We've seen ketchup displaced by salsa and we'll see our presidential candidates speaking Spanish.

The Coming Years

Can the planet survive its huge increase in population? Yes, but many people will die of starvation, yet the human race is infinitely resourceful so we will also hope for vast improvements like those of the past 100 years of incredible changes. Nothing is inevitable but with people living longer, choosing the number and spacing of their children and being more satisfied with their work, we may hope for a bright future.

The United States population will probably grow to some 400 million by the middle of the century, but 80% of that growth will come from immigration. The U.S. continues to admit a large number of people. Where will they live? What will they do? What will they consume and produce? Similar changes of rapid growth are predicted for Africa and South Asia.

People can argue about the numbers, but the point is we're going have to deal with them and provide them with opportunities if we're going to avoid having more conflict and more human misery. The challenge for the international community and those concerned with population issues is to help countries serve their citizens so that they can make their lives better during the century.

A hundred years ago, there was great concern about what to do with all the horses and the manure that was polluting our cities. The horses were displaced by the automobile. Too many automobiles? Perhaps mass transit is a solution. I come to work by electric train. We have solar panels and all sorts of new technology coming about. Society will get in balance, one way or another. It might be a painful transition, but we need to design policies and promote programs that will be most beneficial for the most number of people.

United Nations, Department of Economic and Social Affairs, Population Division, [2 United Nations Plaza, Room DC2-1950, New York, NY 10017, tel: (212) 963-3179, fax: (212) 963-2147, www.un.org/esa/population/unpop.htm].

Margaret Neuse, Office of Population/Reproductive Health, USAID

The Office of Population/Reproductive Health at the United States Agency for International Development provides support not only in the areas of population and reproductive health, but also in HIV/AIDS, maternal health and child survival. Some 30 central program officers manage the 45 staffed missions located around the world. USAID analyzes each country to determine budget projections to alleviate unmet needs and assesses the delivery infrastructure in each host country to determine which channels will offer the greatest efficiency – the government, NGOs or social marketing.

The Population Environment Fellows Program has placed some 70 graduates with degrees in public health in developing countries to work on population and environmental links. USAID has also begun to collaborate with local environmental organizations focused on protecting endangered biodiversity regions. Surveys by these in-country groups of those encroaching on these regions reveal that health is a major concern and this has, in turn, linked the environmental organizations with local family-planning organizations to partner in the delivery of services that are aimed at preserving these endangered ecosystems.

Margaret Neuse is the office director of the office of population/reproductive health at USAID.

Contraceptive Prevalence

The success stories in Africa and Asia have taught us some critical lessons. Following the economic shocks in Indonesia, contraceptive prevalence seemed to hold steady, in spite of the fact it's expensive to use family planning and people had to go to the private sector to get it. Many, recognizing the benefits of such, were willing to pay what they needed to in order to obtain better quality healthcare. That happens when people feel they are in a position of knowledge and can demand such services at the community level. It can happen, in part, from mass media, as it did in Kenya, Zimbabwe and Rwanda before the genocide. There you had community-based workers promoting and informing people about family planning, letting them know it was something reasonable they could do and that they had access to these available services.

We have maintained a critical focus on family planning as our core business. We now describe our initiative as "population and reproduction health," but it is really all those things that connect clearly with the family-planning agenda. We are trying to figure out now what is the next generation of endeavors and how can we be even more flexible. There are some big questions out there.

Recently we've begun working with organizations and communities in areas surrounding biodiversity hotspots that usually are protected in some way – such as parks or nature preserves. In those buffer zones, frequently, large communities tend to be looking for new areas to exploit or farm. So how do we best work with those communities to persuade them to protect the adjacent resource? What kind of assistance do these communities need?

Health is always high on their priorities list, including family planning. Some of the conservation organizations have found themselves in an odd position of needing to know something about how to do family-planning services because these are directly linked to protecting the environment. In some places they've linked up with local family-planning groups or even hired people to provide these services. Right now in Madagascar we have a mission there focused on population-environmental linkages. Madagascar has an incredibly rich biodiversity, but a large, growing population with a huge unmet need for family planning. It all comes together there.

Globally, population growth has slowed. That doesn't mean it has stopped and the projections are still very large for some parts of the world. The assumption in all these projections is that services will continue–which results in an inevitable, downward decline in population growth–but that's not what is happening. It seems you can plateau well above replacement-level fertility and that a decline is not automatic.

A U.S. National Security Council paper recently laid out some global concerns regarding terrorism and an item cited was population and the numbers of young people in particular countries. The hotspots for terrorism tend to relate to the areas of rapid growth in young populations. All you have to do is look at our country. One of the reasons our crime rate has gone down is that the percentage of young people is reduced. A certain age group commits most high crime. It is vital that young people have a sense of hope and some control of their future.

Thus in Kenya, for example, there was an incredible change in attitude regarding desired family size during the 1980s when it went from about six to three in a mere decade–making it one of the fastest declines ever seen in the world. A contributing factor in Kenya was that parents wanted to send their children to school because they saw education as the way to get them a future. Since schooling is expensive and requires fees, uniforms and books, parents had to figure out how to reduce their family size so that they could economically accommodate their educational goals for the family. Parents conveyed to field workers throughout the country, "I want to send my children to school."

The minute people see a future for themselves, family planning is one of those liberating routes to a future. It is an incredible gift to provide the means that allows parents to decide when and how many children to have.

usaid, [1300 Pennsylvania Avenue NW, Washington, DC 20523,
tel: (202)712-4810, fax: (202) 216-3524, www.usaid.gov].

Duff Gillespie, U.S. Agency for International Development

Now a senior scholar with the Bill and Melinda Gates Institute for Population and Reproductive Health at the Johns Hopkins Bloomberg School of Public Health and a public health professor with the school's department of population and family health sciences, Dr. Gillespie will continue efforts he began at the Packard Foundation to increase commitment to reproductive health, child survival, HIV/AIDS, maternal health and nutrition through discourse among globally engaged policymakers. He recently developed an initiative with the World Health Organization, USAID and the Packard, Hewlett and Gates foundations to incorporate family planning into prevention-of-mother-to-child transmission programs, authoring papers and making numerous presentations on the topic.

Dr. Gillespie was the senior deputy assistant administrator of the global health bureau at the U.S. Agency for International Development (USAID) until he retired from government service in December 2002. While there, he played a leading role overseeing a global program in population, health and nutrition with a total budget of $1.7 billion, working in 56 countries. He has received numerous awards for his superior public service.

U.S. Family Planning Programs and the Rest of the World

The United States still has the largest family-planning program of any donor and the level of financial support for that program has actually been quite stable. Among most of the other foreign donors, particularly the Europeans, commitment to family planning in terms of funds has actually gone down. I used to hope the political firestorm that seems to have enveloped the U.S. family-planning program (a small program considering the entire federal budget – somewhere in the neighborhood of 400 million dollars a year) would die down. I'm convinced now this will never happen.

There are some positive changes occurring. The declining and aging population in most of Europe helps, but you still have countries with population growth rates over 2% which means their population is going to double in 25 years. Such trends will exacerbate the relationship of the haves and the have-nots and mean large populations will migrate between countries. It will surely force the developed world to take into consideration the plight of those in the developing world.

Even within the developing world you have a tremendous gap between the better-off and the poor. In Bolivia, for example, contraceptive use amongst the richest 20% in that country is about 40%, but among the poorest there it is less than 10%. That's a huge gap and we're not getting services or information to those most in need of them.

What you can do in Africa is going to be different than what occurs in Asia or in Latin America. There are successful projects in Africa, but it's unrealistic to envision sudden, dramatic changes on that continent.

The Bush Administration

The existing Bush administration is rather complex. In terms of USAID, the support for family planning and the state department, of which USAID is part, has remained strong. There is firm staff and financial support and our various missions have viable programs, but the policy and pronouncements of other parts of the government have caused confusion and had a chilling effect not just on the issue of abortion, but just the appropriateness of contraceptives, whether or not we should have abstinence-only programs, excluding family planning, and whether or not we should have programs working with commercial sex workers to prevent the transmission of HIV. There's considerable angst and apprehension about the U.S. government's support in these areas.

On the other hand, European donors say all the right things, but actually don't have such strong programs; thus, their situation is difficult to grasp. The current U.S. administration comes across as being anti-family planning and it wants to be characterized as anti-abortion. The Europeans claim to be very pro-reproductive rights, reproductive health, but they're not putting the money where their mouth is. The U.S. is funding the projects.

The president of the U.S. has said that one of the best ways to prevent abortion is to have family planning and that we need to have postabortion care programs. Not everybody in the administration, including those in senior positions, agrees with that. There's a debate going on and certainly some things have happened under this administration, and continue to happen, which are very unfavorable for the rights of women.

At one level it seems this administration would like to see the whole issue of family planning disappear. At the most senior levels, I don't think there is a strong commitment on either side of the issue. Unfortunately family planning has now gotten caught up in domestic politics and both sides have dug their heels in deeply and have great mistrust for each other. Such debates actually don't exist overseas, so in some ways we've exported this battle abroad.

There are a handful of countries where it's an issue at a much lower level of intensity–like Guatemala, parts of India and the Philippines. But you can cite them on one hand. This debate is not one of the United States' prouder moments because we are exposing our differences in a way that is unnecessarily imposing our belief systems on the rest of the world. Few donors are engaged in this debate. If any one of them were to drop out, there would be a major problem–like the United Kingdom, Netherlands, Germany, UNFPA and USAID.

Although the United States is the largest donor, this is not a very large foundation on which to base a contraceptive security system. So we have to expand that base and get those countries that can pay for their contraceptives to do so, such as the Philippines, which has never paid for its contraceptives in the 40 years they've had a population program. They can afford it; they should start paying for it.

For those people in countries that can afford to purchase contraceptives, they should take fiscal responsibility. Those who can't, governments and donors need to provide them with the information and services so that they can fulfill their fertility regulation desires. That's not happening, and we're not going to be able to do that unless we get sufficient supplies in the pipeline. The challenge is simply trying to stay on top of the existing population momentum – this huge number of people in the reproductive age throughout the world – which is a great concern. All told, there are a lot of twists and turns in international family planning, but the overall trend is a negative trend.

Duff Gillespie, Ph.D., The Bill and Melinda Gates Institute, Bloomberg School of Public Health, The Johns Hopkins University, [615 N. Wolfe Street, Room W4509, Baltimore, MD 21205, tel: (410) 502-0696, fax: (410) 955-0792, e-mail: dgillesp@jhsph.edu].

Don Weeden, Weeden Foundation

The Weeden Foundation has embraced the protection of biodiversity as its main priority. Population growth (particularly in the U.S.) and over-consumption have evolved into major program interests in order to address more fully the factors driving biological impoverishment. The Weeden Foundation supports organizations that protect ecosystems and wildlife as well as those that increase awareness about family planning. It has financed the first debt-for-nature swap protecting the Beni Biosphere Reserve in Bolivia and is particularly interested in new and innovative efforts to help develop sustainable models for conservation action.

Don A. Weeden, executive director of the foundation, prior to this worked with Ipas in Asia. He hopes at the Weeden

Foundation to continue his grandfather's and his father's legacy of preserving wilderness areas for future generations.

Ecological Footprints

Started in 1963 by my grandfather Frank Weeden, the foundation has from the start focused on protecting biodiversity. He lived in California his entire life and saw California go from having virtually no population at the turn of the 19th century when he was born to today's burgeoning population of more than 35 million. Both population growth and the overuse of natural resources and their effect on biodiversity concerned him deeply. Most Americans are unaware that our population growth rate is fairly substantial, particularly in states like California which is growing at 2% annual growth – higher than the rate in Bangladesh. California's population is slated to go to 58 million in the next 40 years.

Initially the foundation did not focus on immigration as a concern, but increasingly we've seen it as a major factor in the ever-growing U.S. population. This past decade we've begun funding efforts to reform our current immigration policies, feeling we need to focus more on the consumption side of the equation and America's ecological footprint – which is the estimate of how much land is needed to provide the resources and handle the waste products of a typical American. Our ecological footprint in the United States is extremely large.

A number of groups are working on trying to change consumption attitudes in the U.S. There's a group that is creating alternative ways to celebrate Christmas and a group that has created a "buy nothing day." They have been promoting this for about ten years to take place the day after Thanksgiving, traditionally the largest shopping day in the U.S. It's basically looking at this issue of consumption: Do we really need to keep buying and consuming? Economists say it's almost a patriotic duty now, but is it really doing our country long-term good? Many of us feel this kind of consumption is not sustainable.

At times I feel guilty taking long trips, but there's now a worldwide travel agency that allows you to purchase a carbon offset every time you buy an airline ticket. You pay for the additional cost to offset the carbon burned during your trip and this goes towards planting trees or some other offsetting means. I know airplane travel is a problem for those of us working in international conservation development and population, but this is one thing that you can do.

The Joy of Doing What You Believe In

It's a privilege to work at a foundation or a nonprofit where every morning you get up glad you've got a purpose. I care about the environment deeply. I was fortunate growing up in that I was able to get into wilderness areas. I went on a fantastic family trip with my father, Alan Weeden. We spent a full year visiting wilderness gems around the world–and did it on a shoestring, camping out 200 nights during that year. This experience cemented my deep appreciation for the spiritual value of wilderness. I've come to realize in working in this field how important it is in terms of our health as well. Our economy is already seeing the effects of over-exploitation of natural resources and the diminishment of natural habitats, not to mention the loss of biodiversity, which is really a crime. Our generation and the ones that went before us are going to be remembered largely for the negative impact we've had on the earth.

The Nepal Experience

Our current population of over 6 billion people is placing a great deal of pressure on natural habitats all over the world. It's not just multi-national commerce, but the pressures are basically derived from over-consumption in the U.S and other parts of the world. A lot of it is local consumption. I spent several years living in Nepal where the population/environment equation really became apparent to me.

At first sight Nepal is a bucolic, lush environment when you're in the mountains. Initially it seems as if they are living in harmony with their surroundings, but soon you realize the farmers have been pushing beyond the sustainable limit, up steep hillsides in the quest for new land to till and that the per capita farmland, as a result of the population growth, has been decreasing in the hills over the past several decades.

There is also a continuing pattern of what is known as out-migration. The result is a patchwork of environmental fragments. Only a few areas left throughout the Himalayas retain any natural ancient forest. Those fragments are largely confined to the eastern Himalayas because next door in Bhutan they have managed to preserve most of their biodiversity mainly because of their low population growth. In Nepal the impact of humans on the land is considerable–they're consumption farmers using low technology that impacts the land directly with erosion, siltation of rivers, massive deforestation and a cycle of environmental degradation.

Similar syndromes to Nepal's are happening all over the world. Thinking of a 50% increase in global population, almost entirely in developing countries, the mind boggles. Poor farmers have few options, that is why in Nepal, you either try to farm higher up the mountainside or you try to move someplace else because in Nepal there is no more land left to move to.

The Weeden Foundation, [747 Third Avenue, 34th Floor, New York, NY 10017, tel: (212) 888-1672, fax: (212) 888-1354, e-mail: weedenfdn@weedenfdn.org, www.weedenfdn.org].

Dr. Carmen Barroso, MacArthur Foundation

Prior to assuming her current position as the regional director of the IPPF/Western Hemisphere Region, Dr. Carmen Barroso served as director of population programs for the MacArthur

Foundation which had entered the population and reproductive-health areas in the late 1980s. Their initial goals were to highlight population issues in the international development agenda; to raise the issue of women's empowerment through population policies; and to let the voices of the Third World be heard in the international policy arena.

This strategy was focused on countries that were leaders in their respective regions: Brazil, Mexico, Nigeria and India. Dr. Barroso's challenges at the IPPF/WHR include meeting the increased need for contraceptive services, HIV prevention, male involvement, combating gender-based violence and funding challenges based on the Global Gag rule, which IPPF declined to sign and which, as a result, will make them ineligible to receive USAID monies.

The Role of NGOs

In Mexico we fund several networks including one that is promoting sex education in public schools. We have been very effective in getting government approval for the sex-education curriculum and having them publish the guidelines and basic textbooks for teaching sexuality in the schools. This was a major achievement of Mexican NGOs and we are glad we were able to fund some of their efforts.

In spite of a challenging environment, Nigerian NGOs are also doing a great job. We fund a partnership for safe motherhood – a big problem in Nigeria where the maternal mortality rate is huge – and a partnership between gynecologists, women's organizations and other grassroots organizations to lobby the government to provide safe services for childbirth and prenatal care. Because of free services that these NGOs pushed the government to provide and the government's involvement in sex education, the maternal mortality rate in northern Nigeria decreased. Because there are numerous misconceptions about sex education, NGOs must educate the public also. There is fear that if you teach sex in schools, children will initiate sexuality earlier or not respect their parents or will become promiscuous. Research shows this is not the case.

When an NGO enters the field of sex education, it has to develop not only the technical expertise to give youngsters the information they need and the opportunities for discussions they need in order to create their own values and assert themselves, but the NGO also needs to educate the public to accept it as a valuable way to contribute to a healthy environment.

We are glad that we have been very well received. The Bush Administration's policy of Abstinence Only is not realistic. We should encourage abstinence to youngsters when they are not mature enough for sex, but we should be realistic enough to understand that in this day and age, children need to know how to protect themselves so that they can have safe sex if and when they decide to initiate their sexual behavior.

Since the 1970s Brazil has had a strong array of women's organizations and an incredibly rapid decline in its total fertility rate. The women's movement pushed for policies that let women decide for themselves what they want, so it was powerfully effective in terms of changing the mentality of the nation. Even poor women want their children to have a better chance in life and they realize that for that to happen children need to go to school. Also as increasing numbers of women have to work, they know by having a large family, their options for working outside the home are restricted. So there is a direct competition between the need to work and the size of the family. Research has repeatedly shown that women who work outside the home have fewer children.

The Challenge of Safeguarding Women's Reproductive Rights

The MacArthur Foundation has fairly large population/environmental and women's reproductive health programs. We are facing not only needs for contraceptive services but also huge needs for HIV prevention and we are trying to get male involvement in combating gender-based violence. That is a political challenge because right-wing extremists, who have powerful friends in the U.S. administration, are increasingly active and are threatening all our programs. This affects our funding, because with the economy hurting, several foundations are either reducing their programs or changing fields, so we need to face these challenges.

The Global Gag Rule is absurd because it curtails the freedom of expression of the many organizations working to save women's lives. If you receive funds from USAID for providing contraceptive services, you are not allowed to discuss your views about abortion issues. Clandestine, unsafe abortions are one of the main causes of maternal mortality. Dr. Mahmoud Fatallah (a consultant with the International Federation of Gynecology and Obstetrics, FIGO) says these women are dying not because we don't know how to save their lives since we certainly have the knowledge of providing safe abortion, but because society has decided their lives are not worth saving. That summarizes the effect of unsafe abortion.

In my early days I used to have monthly discussions with Catholic bishops. The press loved these public discourses and I learned a lot from those sessions. I respect their position. I only wish they would respect ours. I think if somebody believes life begins at conception and that this life is as precious as a woman's life, that's their privilege and they should act according to their beliefs. But if we think that women's lives are important and that they have the right to decide whether or not to have children, they should not be coerced and they should be respected as well.

International Planned Parenthood Federation, [120 Wall Street, 9th Floor, New York, NY 10005, tel: (212) 248-6400, fax: (212) 248-4221, e-mail: info@ippfwhr.org, www.ippfwhr.org].

Ana Maria Goldani, University of California, Los Angeles

Brazil has a low birthrate because there has been rapid decline in infant and young-adult mortality, an increase in women's education and through their participation in the work force, women have been given greater equality and empowerment, children are now very expensive to raise, currently sterilization is legal and family-planning choices increasingly available.

The beginning of fertility decline in Brazil started with the over-the-counter availability of oral contraceptives in the 1960s and then accelerated with the availability of female surgical sterilization. From 1970 when the fertility rate was close to 6.2 it dropped to replacement level by 2000. From 1986 to 1999 the percentage of women in the reproductive-age group having sterilizations grew from 24 to 43. Before the recent change in sterilization law, C-sections, most of which were unnecessary, were performed in conjunction with sterilization and were often an excuse to get sterilization. Of the estimated one million illegal abortions performed each year in Brazil, over a quarter end up with medical complications putting women in public hospitals. Therefore, providing safe, post-abortion care is a constant challenge for the medical community.

Throughout Brazil and especially these past few years in the northeast, there has been improved access to health services so infant mortality has dropped dramatically. Amazingly, the public-health sector is now more actively involved in delivering family-planning services and responding to a large number of women's organizations that are demanding equality and justice. The Catholic church is equally concerned with preventing infant mortality and maternal mortality and morbidity.

However, the needs and guarantees of the reproductive rights of millions of Brazilian women are not being met and Brazil faces about a 15% increase in the size of the reproductive-age population in the coming two decades. There is growing concern that supplies of reproductive health commodities for family planning, safe motherhood, prevention and treatment of sexually transmitted diseases will become even scarcer due to the economics and inequalities of Brazil and the reductions in international aid.

Ana Maria Goldani, Visiting Associate Professor, Department of Sociology, UCLA, [264 Haines Hall, 155103, Los Angeles, CA].

Dr. Joseph Speidel, University of California, San Francisco

Dr. Joseph Speidel served as the acting director at USAID from 1979 to 1983 where he was one of the creators of the World Fertility Survey, now known as Demographic and Health Surveys (DHS), which measures in different countries the attitudes, knowledge and use of contraception and desired family sizes. After USAID, Dr. Speidel became president of the Population Crisis Committee (now Population Action International) and a program officer of population issues at the Hewlett Foundation. Currently he works as an adjunct professor with the University of California at San Francisco on public health issues.

Basic Points

First, since we know how to help people have the desired number and timing of their childbearing, we should help them do that. The United Nation has stated this is a basic human right. Secondly, it's very critical to health. Reproductive health is one of the most important health issues we have, especially with AIDS today, but also death in childbirth. A recent study by the Global Health Council documented approximately 300 million unintended pregnancies in the past five years with 700,000 deaths relating to unsafe abortion and, even more importantly, to unsafe childbirth. Third, rapid population growth impedes the social and economic progress desperately needed in the developing world where most of the earth's population lives and where 98% of the population growth is occurring. Finally, rapid population growth and burgeoning populations put stress on the environment. We cannot add a billion people to this planet every dozen years or so and preserve a sound environment–which is important intrinsically, but also important to our own welfare.

Something like 44% of the world lives in low fertility countries. Even those countries, China being one of them, continue to grow demographically. With around 1.2 billion people, China is adding ten million people each year because of the age structure. Momentum is the number one reason population is growing. Sadly, about half of the pregnancies every year, including those in the U.S., are unintended–either mistimed or just not wanted. Collectively the world is not doing a very good job here. We could get halfway down to a two-child family if we just resolved the crisis of unintended pregnancies. The approximately 36 million abortions every year that result from unintended pregnancies could be avoided.

13 Billion People?

With 44% of the world's people living in countries with low fertility, close to that 2.1 average number of children needed for population stabilization, we have a big success story, because some of these countries are developing countries. It's the 56% of the world living in rapidly growing nations that present the problems. Were today's fertility to persist over the next 50 years, we would end up with 13 billion people, over double today's total. The United Nations predicts we'll only get to 8.9 billion, but to make their projection come true, we need to do a lot of work.

Couples and individuals all over the world want to control their fertility. Generally they want smaller families. By getting information and services to them–including abortion services if contraception fails–the world's fertility problem would be solved. This is a job we know how to do. We just need the political will and financial resources to make it happen.

The University of California at San Francisco, UCSF, has a strong commitment to four general areas: work with adolescents; work on abortion; work with sexually transmitted infections and AIDS; and work on family planning and contraceptive development. There's an enormously successful program in California to bring family planning to the poor. Throughout the state the unmet need has dropped from 70% down to 37% over four years with 1.2 million women, men and couples now being serviced. The rates of unintended pregnancy, abortion and teen pregnancy are all dropping. We need to get that message out to 49 other states.

There's a presumption abroad that since we've made a lot of progress, the problem is solved. Scrutinizing the data shows we have a long way to go. Because there has been high fertility in the past, this means the numbers of couples of reproductive age are growing rapidly. Just to stay even, we have to do more. This causes controversy and makes people nervous, for unfortunately, unlike general health which everybody is for, a lot of people are against population work, against family planning and especially against critical areas like making safe abortions available. This puts us under enormous handicaps, but if one thinks it through and considers the big picture – how we can assure social and economic welfare around the planet – then you have no choice but to remain engaged with population.

The links between population and environment are so strong I'm surprised more environmental organizations and people who care about the environment don't get involved in the population issue. Whether you are in a developing country where per capita demands on the environment may be less, but the numbers of people are huge, or in the developed world where consumption patterns are extremely high, there's over-population in both places. We have this biological bind that we're in. The natural-resource base is shrinking while the population base is going up. This is not a pretty picture.

Center for Reproductive Health Research & Policy, [3333 California Street, Suite 335, Box 0744, San Francisco, CA 94143-0744, Tel: (415) 502-4086, fax: (415) 502-8479].

Dr. Paul Ehrlich, Stanford University

Best known as the author of *The Population Bomb,* Paul Ehrlich was cofounder with Peter H. Raven of the field of co-evolution and has pursued long-term studies of the structure, dynamics and genetics of natural butterfly populations. He has also been a pioneer in alerting the public to the problems of overpopulation and in raising its impact on resources and the environment as matters of public policy.

Professor Ehrlich's research on the dynamics and genetics of natural populations of checkerspot butterflies (*Euphydryas*) has applications on such problems as the control of insect pests and optimum designs for nature reserves. He has also led investigations in ways that human-disturbed landscapes can be made more hospitable to biodiversity in what has been called "countryside biogeography," now under the direction of Dr. Gretchen Daily, founder of the field. The Ehrlich group's policy research on the population-resource-environment crisis takes a broad overview of the world situation, but also works intensively in areas of immediate legislative interests like endangered species, the preservation of genetic resources and cultural evolution – especially with respect to environmental ethics. He has been much applauded around the world for his work.

The Population Bomb

Since I wrote *The Population Bomb*, more people have been added to the human population than were alive when I was born. The book helped people focus on the fact that one of our major drivers in environmental deterioration was growth of the human population at an extremely rapid rate. The first thing I would say to world leaders is we're already vastly over-populated. You can see this because we're not living on our income; we're living on our biological capital which is something we can't do forever. It's not just a matter of how rapidly or how large the population is, but how individuals behave. By those standards, of course, the United States is the most overpopulated country on the planet.

My colleagues and I have been totally unsuccessful in changing consumption patterns. You can see it in the current U.S. government which has just fought a war basically over the narcotic that our society's hooked on – oil. It is unlikely we would have invaded Iraq if their main natural resource and what they had the biggest supply of was broccoli. The main thing everyone must remember is they've got to get involved politically and pay attention to what's going on politically.

Another big shift needed in the foundation community is to focus on who's doing a great job and then support them without making more trouble for them. In the scientific world we are trying to change the basic ethics of the entire community. If we recycled everything we possibly could and kept consumption and population growing, the recycling might give us an additional day before we went down the drain. That's not where the action is.

A massive assemblage of scientists got out a "Warning to Humanity" in 1993 sponsored by the Union of Concerned Scientists and signed by more than half of the living Nobel laureates – 1,500 distinguished scientists basically saying if we don't change our ways, we are doomed. Even with public relations firms helping out, they couldn't get a mention in the *New York Times* or in the *Washington Post*. It just wasn't deemed news. That same year, 50 academies of science around the world got out a statement on population and environment, which said basically the same thing. That never broke through into the media either.

Biodiversity

Probably the most important part of the capital that we're spending and getting rid of is our biodiversity – the plants, animals and microorganisms which as far as we know are our only living companions in the universe and which, more important from a human point of view, are the working parts of our life-support systems. We're sawing off the limb we're sitting on. We have entrained the sixth great extinctions of other forms of life on this planet. The pandas can disappear and many won't care, but when we're destroying the organisms that supply the ecosystem that provide us with the fresh water that maintains the gaseous quality of the atmosphere without which we couldn't grow crops because they'd be destroyed by pests that generate and maintain the soils that are absolutely essential to keeping us eating, we're getting rid of ourselves at the same time.

I don't think there's more than a couple people in Congress who could make that speech to you. Most of them are unconcerned with the fact that the very underpinnings of our economy and our civilization are being destroyed as we wipe out biodiversity. If you hear a politician say we've got to consider economics before we consider ecology, it's like saying, you've got to consider eating before you consider breathing. If you stop breathing, who cares? There will be no economy if the ecological systems are not maintained. We're not maintaining them; we're destroying them ad lib.

I've been doing fieldwork for over 50 years watching population after population of the organisms that I work on disappear. It bothers me to see lots of beautiful organisms needlessly wiped out because when you wipe out those organisms, what really bothers me is you're cutting off my grandchildren's options and future.

Immigration

The fact that many environmental and population groups don't want to tackle immigration is because it's an extremely complex and difficult issue and they're afraid they'll lose members if they do so. The immigration solution is really very simple if you look at it from a scientific point of view. Those who want to maintain a quality of life in the United States must understand we have way too many people today and we've got to gradually reduce the numbers. You can do this either by having fewer births or by having fewer immigrants.

Every time you allow in an immigrant there should be a trade-off for a birth. How you balance births and immigrants is a social issue that ought to be discussed–but you've got to have that balance. It's insane to discuss how many people you let come into the country without talking about how many there are going to be in it. Immigration's a tough issue. When people come to the United States they tend to adopt the American consumption patterns which is the very worst thing: more and more super-consumers.

What the Future Holds?

Some people take the stance that another billion is only an increase of 15-20%, whatever it is. The planet doesn't compute percentages. Another billion people will put enormous stress on our life-support systems and the next billion will put disproportionately more because human beings farm the best lands first, get the most accessible water first so that every additional person means you've got to haul water further, there's more chance of it being polluted and you've got to go further for your other resources. The general view of ecologists is stop as soon as you possibly can and start a slow decline because there's no guarantee we can get over the hump without a tremendous disaster.

A disaster, for example, could be the deaths of several billion people from an epidemic worse than the 1918 flu. With today's level of crowding, one of the biggest worries is a new epidemic we can't control. Another huge disaster would be a nuclear war over resources. If we had a sudden climate change that terribly disrupted agriculture it would be quite possible for billions of people to starve to death. There are lots of unpleasant possibilities compared with simply limiting our births to the point where we start a slow decline. It's impossible to know where the point of no return is.

My greatest fear is we've already passed it, but I see no choice but to pretend we haven't and do the best we can to ameliorate and cut the suffering. Right now we are already not taking care of fully half the number of people that were alive when I was born.

The most powerful country in the world is in a 1933 Hitler's Germany sort of situation. Civil liberties are disappearing, women are being attacked on every front and the environment is being destroyed. Fortunately, human societies can change rapidly. When I was a child they were still lynching African-Americans in the South, holding up little girls to see it, taking home pieces of flesh as souvenirs. In a very short time, during the 1950s and 1960s, the whole racial situation turned around. If I had told you in 1985 that in a decade the Soviet Union would be gone, Communism would be on the run, everybody realizing it was a failure and the Berlin Wall down, you'd wonder what I was smoking. Our job now is to try and ripen the time for the kinds of changes we need to treat each other and our environments properly so that we all survive.

I say there is hope that we can change, otherwise I wouldn't be wasting my time with you. I'd be living on Bora-Bora in the middle of my wine cellar.

Bob Gillespie's Response to Paul Ehrlich's Discussion

Paul Ehrlich started his career as a field biologist and distinguished himself working with butterflies and publishing books on ecosystems. The population of the planet has almost doubled since he wrote *The Population Bomb.* He's continued to enlighten politicians and the media and has spent his career looking at ecosystems being undermined by nonsustainable consumption levels. The mathematical models he's designed allow us to see within the framework of the technology, population and resources what is a sustainable future.

Paul is telling us we may have already passed the fail-safe point of irreversible ecosystem decline. You cannot bring back the thousands of species lost due to human overpopulation or over-consumption. Once those species are gone, the only thing we can preserve is genetic material. His career has been dedicated to preserving and protecting the environment, trying to create an awareness of the population explosion and internalizing individual commitment to small families – which is more than stopping at one or two children; it's living sustainably.

His book *Healing the Planet* outlines specific actions politicians, economists, journalists, scientists and engineers need to take to prevent over-consumption and overpopulation. Scientists continuously tell us we are going down a suicidal path, but their reports, statements, declarations and findings are ignored. If we destroy our ecosystems, we destroy ourselves. The problems are going in one direction and the potential for solving those problems are going in the opposite. Paul has been a courageous and outspoken advocate telling us we need to adjust our cities and civilizations to zero-based fossil fuel economies. Much of Ehrlich's writings prescribe the actions required for us to live in a world of hope. He is committed to teaching young scientists to produce a new generation of scientists willing to address these substantive issues – and continues his deep commitment with enthusiasm and passionate concern.

Professor Paul R. Ehrlich, Bing Professor of Population Studies, [Department of Biological Sciences, Stanford University, Stanford, CA 94305-5020, tel: (650) 723-3171, fax: (650) 723-5920, www.Stanford.edu/group/CCB/Staff/paul.htm].

Martha M. Campbell, Center for Entrepreneurship in International Health and Development

Dr. Campbell, codirector of the Center for Entrepreneurship in International Health and Development (CEIHD), founded in 2000 on the campus of UC Berkeley, currently lectures there. The center's mission is to promote and disseminate the use of entrepreneurial methods to improve the health of families in developing countries. Having begun her career in public health at the Planned

Parenthood Federation making access to family planning easier for poor women in the U.S., she subsequently conducted research on competing perspectives on population around the world.

A Population Problem?

Most Americans find it hard to believe there's a population problem. From an airplane, you see the American wide, open spaces and it's hard to think there can be a population problem. Even flying over Italy, Greece and much of Africa, it is not something that is comprehendible. To do so, you must understand the problem of resources, water, growth and the drawdown on water tables, in China for example, where water tables are dropping like a rock. You must understand how limited the resources of the Nile are compared to the population of the countries dependent on it–which in another 50 years will have an additional 350 million people in need of its water, which is inadequate for providing such demands. India is adding by net growth a million people every three weeks. They are resistant to the subject and don't quite believe there's a problem, thinking people will control their family size if they want to. That assumes rational choice works in family size, which it doesn't.

Conflicts, Consumption, Feminism and Male Dominance

At the Rio Summit it was voices from developing countries, not necessarily their leaders who said they were offended and uninterested in talking about population and rather wanted to discuss the consumption of the North–which they saw as the principal cause for environmental decline, not a burgeoning population. Women at the Cairo conference reiterated it wasn't about population. Granted, demography was bad, but their concern was women's health issues and other aspects of development, so they shifted money from family planning into other aspects of healthcare. Women's health and development are vital to the development of the community, but taking attention away from population and family planning does a sad disservice to women in the long run.

The economists were free market business leaders and they see population as no problem because it provides a growing market and labor force. Basically the major criticism against limiting population came from the men who are threatened by women making choices and their desire to remain dominant in society and in the family by controlling women. Though everyone gives lip service to supporting women's reproductive rights and health, many fear that discussing population might lead to coercion. Without family planning, women lose the initial right, which I think is prerequisite to all of the forms of empowerment. Population growth is a real problem in many parts of the world and it is not a popular topic right now.

China's one-child family program followed decades of Mao Tse Tung's lead where women could not get family planning, which produced an explosive population growth. The many critics of their compulsory program do not factor in the sinking water tables and disappearing arable land which is incapable of supporting the population growth rate they had. Those who criticize China's population program tend not to understand the scale nor the challenges of China's environmental crisis.

Religion

If the woman can't decide, who is supposed to make that decision and why should somebody else do it? What is it about some religions that makes them punitive on women? I don't understand why some religions are so adamantly against abortion claiming this is the "word of God." It makes me sad that so many women have to suffer and die because of decisions made in the name of religions. Religion is supposed to be about kindness and care. I'm all in favor of those who are against abortion not having abortions. I feel they should live by their own principles, but when they support rules mandating everybody else to have unwanted children, I think that's totally inappropriate, totally unfair and unkind.

Martha Campbell, Ph.D., co-director, Center for Entrepreneurship in International Health & Development, UC Berkeley, [140 Earl Warren Hall, Berkeley, CA 94720, tel: (510) 643-2700, e-mail: campbel_mm@yahoo.com, www.ceihd.berkeley.edu].

Dr. Malcolm Potts, School of Public Health, UC Berkeley

Dr. Malcolm Potts' chief areas of research are cost and mobilization of resources for international family planning, AIDS-prevention strategies and resource needs and the biology of human sexual behavior. He thinks human sexuality must be treated as a biological act, not a moral action. Until the 20th century, the average onset age of menses for young women in America was 18. With the improvement of diet that led to higher body weight for Americans, the onset of menses has dropped six months every decade in the 20th century. In preliterate cultures such as New Guinea, women have their first child in their early 20s and most have never seen modern contraception. Early puberty creates a biological catastrophe for the advent of teenage pregnancy increases risks for young women whose bodies are not ready to bear children plus population growth in societies with early onset of puberty.

The Perceptions of an Embryologist

I was a young obstetrician in England before the abortion law was changed, so I saw the misery that women endured in trying to control their family size and started offering women family planning. Personally, I think abortion is a deeply ethical problem, but when life begins is not. As an embryologist, I think it's easy to exclude the extremes. I don't think that a fertilized egg is the same as a newborn baby. I, and every woman on this planet, can distinguish between those two. I respect those who differ in their belief about when life begins, but in a pluralistic society such as we have, we need to be able to live with different points of view, just as we have to live with different interpretations of life after death–which Muslims, Jews, Catholics and Protestants all have very different perceptions and analyses.

Economics and Family Planning: A Credo For Success

We're the largest economy in the world–and possibly the largest contributor to international family planning. In relation to our population and wealth, we're one of the meanest–far behind many of the Europeans and the Japanese who are the largest contributors to international aid in the world. If we could somehow shape our tax system to take a nickel per day from each taxpaying citizen of the United States, a country that expends a billion dollars a day on defense, those few pennies would add up to a few billion dollars a year, and that would make the world a more secure place in a generation's time, which is one of the things that family planning does.

In my lifetime I've seen some wonderful successes. Thai couples went from six children to two in about 30 years; Koreans from six to 1.8 in that same time. Iranians saw their fertility rate decline from 5.5 to two in a mere decade. All these countries basically did the same thing: they respected people's choices and gave them a variety of services. There's no one-shot solution. Through a multiplicity of channels of distribution, people could get needed supplies from clinics, pharmacies, even in various shops, and they recognized the role of voluntary sterilization–the number one choice in the United States, by the way, as well as in many countries. All those nations that have succeeded at family planning have also recognized the problem of abortion, in the sense that they were pragmatic and realistic about the fact abortions happen, and that they are better done safely.

Being realistic means recognizing that contraceptives don't

always work. On the island of Bali in Indonesia one physician offered termination of pregnancy to women who got pregnant with an IUD if that's what they wanted. This option made family planning there extraordinarily successful and Bali was the first region in Indonesia to have a two-child family. We must acknowledge that of all the women now entering their fertile years, every 15-year old woman on this planet, on average will have one induced abortion before she reaches menopause. This is not unusual or rare. We should strive to make it as unusual as possible by offering good contraception, but for the foreseeable future, abortion is going to remain an important variable from a woman's point of view and from a health point of view. It is something that is part of the family-planning equation, let's admit it.

Family planning in Korea, Taiwan, Sri Lanka and some parts of India – not the north, but the south – have been very effective. Mexico's program commenced somewhat late, relative to these other nations, but has done very well. Brazil's moved more rapidly than I expected. There were very important failures which is producing some worrying divisions in this world.

The low fertility countries will move forward economically, become politically stable and inevitably more democratic simply by way of the information revolution. That leaves us with high fertility countries where successful family planning is not offered who have terrible problems. Nigeria, Pakistan and Afghanistan could suffer terribly as denizens of those regions endure totally unacceptable unemployment rates which, in turn, make for politically unstable governments. This is precisely where so many terrorists come from. Osama Bin Laden is the 17th child of a man with 11 wives and 55 children. I suspect not all of those 55 children were wanted.

Wars are complicated and have no simple explanations. Many people, including some working for the U.S. Central Intelligence Agency, recognize what is called the youth bulge – very rapid population growth resulting in a huge proportion of the population being in the 15-to-29-year-old age group. Some are angry, testosterone-driven young men, especially those with little education and no chance of gainful employment. We are spending a lot of money on bioterrorism and homeland security, but what we're failing to do is employ the obvious long-term strategy–offering women choices all over the world, especially in those currently unstable countries like Pakistan.

We know those women want smaller families. If we could simply meet their unmet need for family planning (proven from excellent surveys), they would have an average of two children immediately. There would still be some societies that would take a considerable time to come down, but if family size can go from five to two in Iran in ten years, it can go from five to two in Afghanistan in 20 years, and it can go from five to two in Pakistan in 15 years. That would be the greatest contribution we–Americans, Europeans or any wealthy country–could make to the future stability of this world. Correspondingly, Saudi Arabia could go from a total fertility rate of five to two in 10 years, as well, because it's a wealthy country.

The Media, Controversies, and Health Clinics

Letting the media explore the controversies is healthy. When Pope Paul VI issued the encyclical "Humanae Vitae" that said Catholics could not use so-called artificial methods of contraception, the wise family-planning groups in Latin America like Colombia went out there and used this to say to people, "Hey folks, this is what exists; these are the choices; don't be afraid of controversy; don't back away from it." Using the media and multiple channels of communications gives you multiple channels of distribution.

Often family-planning programs fail because they are too top-down and clinic-oriented. Most women in poor countries have never been to a clinic. Clinics are fine in rich countries for certain groups, but even in America they're not the primary source of health in family planning. Social marketing, where you use the existing infrastructure to advertise, may be much more effective. Every country, even Somalia, has an advertising agency so you can create a brand image for a contraceptive and advertise it through the established wholesale-retail chains. Use what is there, don't try to set up a new structure. Use private doctors and private midwives. It took me a long time in international health to understand that the poorer you are, the more money you spend out of your own pocket on your healthcare. In India the poorest people may get three or four dollars from the government for their healthcare cost, but they're spending twice that out of their own pockets.

The Confessions of a Doctor

Once I diagnose a disease, I can look up the answer. But there's no diagnosis in family planning–it's a set of choices. People misunderstand this and want to make it into a disease they call over-fertility so clever doctors can offer a treatment, like sterilization. That's the wrong approach; rather it should be: As a doctor I have some information I'd like to share with you; I'll answer your questions; you make a choice.

Many aspects of medicine are quite patronizing. For example, the doctor thinks: I know more about their breast cancer than they do, so I'm just trying to get them to take the treatment I think is best for them. But in family planning, it's all about choices. When you offer options, you get a universal response from around the world that huge numbers of people want fewer children. As you meet that need, the people who originally said they wanted four children now opt for three, then they'll say they want two, probably less than two.

We have to understand where we're coming from and be prepared to change our nature in the modern world. We've come from living in small groups of hunters and gatherers to now living in a world of concrete and computers. This requires making changes regarding how the two sexes relate to each other. Interestingly, in family planning, such a change is unnecessary because throughout most of human history, couples had two children. That's very important to grasp. Two children survived to reproduce in the next generation.

Our population explosion is a recent phenomenon resulting from a rapid decline in infant mortality, a change in breast-feeding patterns and much earlier puberty – each of which has contributed to higher fertility. Puberty occurred at 17 or 18 years in the time of Bach. Teenage pregnancy in the 18th century was rare. Early puberty has produced a biological catastrophe. Modern adolescents are torn by the irresistible drives of sex and other recent factors affecting their situation. There will always be mistakes and problems relating to sexual behavior and its outcomes. It is a humane and humble approach to recognize and deal with the fact that young people are in a bind and we should help them, not simply moralize when things go wrong for them. We know early pregnancy and childbearing is not a good thing for anybody – the child, the parents, nor for society. We can help young people reach their goals, notwithstanding the predictability of a lot of mistakes.

The single most important factor in bringing down the age of puberty is body weight. Belly dancers who restrict their diet and exercise a great deal have late or irregular periods like our ancestors. In modern times where food is plentiful, people tend to put on weight more quickly. Other factors contribute as well. Artificial light is suspected as a stimulant, even hot showers. Pheromones, chemicals and environmental conditions also are influences: women living in a dormitory tend to have menstrual cycles together, which is pheromone related. No one knows fully what all affects the onset of puberty.

Do Unto Others as We Would Have Them Do Unto Us

This profound kernel of Christ's advice after he observed the world should inform our decision-making. We have been relying for some time upon the benefits of family planning and access to safe abortion. Our society is built on these foundations. We would not be the prosperous, educated society we are today if we all had six children. If you want to have a large family, that's fine, but people here have been given choices. Those same choices should be extended to all people. In many ways, it's more important to be able to control your fertility than it is to vote or have a free press. Those are great privileges, but controlling your fertility is the foundation on which the other democratic freedoms are built.

We live on a knife's edge today. We could engender an increasingly unstable world which might, for example, see India and Pakistan hurling atomic bombs at each other. Or we might see terrifying bioterrorism occur or other heinous acts too awful to mention. We could mindlessly destroy all the large animals and much of our environment–or, by the end of this century, we could have a fantastic world with our energy problems solved, with most people living a reasonably dignified, fairly healthy life. But we will never go in the right direction without improving our investment and realism about international family planning.

Malcolm Potts, MB, BChir, PhD, FRCOG, Bixby Professor, UC Berkeley, School of Public Health, Maternal and Child Health, [314 Warren Hall, Berkeley, CA 94729-7360, tel: (510) 642-6915, fax: (510) 643-8236, e-mail: potts@socrates.berkeley.edu].

Allan Rosenfield, Mailman School of Public Health, Columbia University

The Mailman School for Public Health at Columbia University collaborates with partners around the globe on merging preventive and treatment strategies and viewing AIDS in the context of other infectious diseases faced by resource-poor nations. It played a major role addressing disaster-preparedness issues raised by the terrorist attacks of 9/11 developing response protocols and training in partnership with local, state and federal agencies.

Recently the school doubled its budget and grant support to develop state-of the-art laboratories for infectious disease research and – by responding to new and emerging public-health crises – has become one of the nation's largest, most respected graduate schools of public health. It has pioneered collaborative project models that employ and empower local public health professionals and community leaders. The diverse student body, the majority being foreign or minority students, graduate to excel in the areas of education, HIV/AIDS, sexually transmitted diseases, globalization of emerging infections, tobacco use, healthcare policy, cancer, reproductive health, violence, environmental health, asthma and clean water issues. Allan Rosenfield, dean of the Mailman School of Public Health at Columbia University, began his professional career as an obstetrician in Nigeria. He has also worked for the Population Council in Thailand.

Title X

We are at a very different place today than we were when I started in this field in the 1960s and 1970s. Contraceptive prevalence worldwide is at 50% where it was only 3% then. What's happened in the last 30 years with family planning is one of the major public-health success stories of the 20th century. No one could have predicted how far we could come by the new millennium – in spite of there being still major unmet needs for family-planning services.

With Title X programs we presently have widespread coverage for family-planning services in the United States, especially for lower-income groups. Initially, some in the African-American community perceived this as a genocidal program and attacked it, but then African-American women told their male partners to cool it because they wanted access to family planning. Today the Title

X program is noncontroversial in terms of reaching minority populations in the United States.

New threats to the success of family planning have been coming from this administration's abstinence-only initiatives. Nonetheless, presently contraceptive prevalence levels are very high in the U.S. We still need to look at how people pay for this. For the past decade I have felt it was justified to make the pill available over-the-counter as we do now with emergency contraception. The problem with this for low-income people, is once it's off prescription, it will no longer be provided in Title X clinics which will make it more costly for them to buy generically over-the-counter. Even though I started lobbying to get the pill off prescription, I revised my stance when I thought about the financing of family-planning services in the U.S., at least until we achieve comparable costs between generic pricing and what the price is at a Title X-funded clinic. There are important fiscal issues for low-income populations in the U.S.

A major focus of mine in developing countries is to take the family-planning experience and integrate it with the way maternity-care services are provided. So when you have no obstetrician or doctor available, can you train auxiliary personnel to insert IUDs or complete unsafe abortion or do a Cesarean section? The issues in America are different – they're not about safe procedures, but getting basic services to people. My advocacy in the United States thus focuses on the 41 million uninsured people who cannot afford healthcare. Title X helps us with family planning, but it doesn't help low-income people get other types of routine healthcare.

Women Worldwide

We can anticipate for some years to come, that the current level of some 600,000 women will die annually from complications of a pregnancy – obstructed labor, hemorrhaging, infection of complications of an unsafe abortion. None of these tragedies requires any new technologies, vaccines or drugs for prevention. We only need to make maternity care services available and accessible to those who need this. A century ago the U.S. had maternal mortality rates at the same level as we see now in developing countries. With the advent of accessible services, antibiotics and safe blood transfusions, we changed pregnancy to a very safe procedure in this country. Our maternal mortality ratio is about 8 deaths per 100,000, compared to Afghanistan, which has the highest anywhere in the world: 1,600 maternal deaths per 100,000 live births, so our program is trying to develop mechanisms allowing women to have access to needed emergency care when they have complications.

The Challenges Ahead

There are countries in Africa right now with as high as 35% of the population HIV positive–extraordinary percentages compared to almost any other disease of the past. Most diseases hit the very young or the very old, but this one hits right in the middle of the most productive years. It's also across the board knocking out everybody: teachers, doctors and policy people and with these levels, you're impacting a huge level of society. Shift to Asia, China and India where there is now 1% or less of the population HIV positive–which is very low. If India were even 10% of its population HIV positive in ten years, more than 100 million people would die from the disease unless we come up with better ways of getting treatment to them.

There are potentially new research avenues looking for better contraceptive methods. Using some of the newer molecular scientific tools we might find something that could impact the process of the maturing of sperm, spermatogenesis as its called, or something focused on ovulation that doesn't have other effects like the pill does. One of the hindrances to advancements has been that the pharmaceutical industry's hesitancy to invest in major contraceptive research. One is litigation. You never know when you develop something that's going to be used by millions of people if there might be an unexpected impact that didn't show up in the clinical trials of a thousand people. The pharmaceutical giants are also concerned about economics. They make a lot of money with oral contraceptives. It takes about 15 cents to manufacture a package of pills. In the United States that package sells for nearly $30 now. That's a pretty good markup. The perfect contraceptive is not going to be a moneymaker. The better they are, the less economic benefit there probably will be.

The nonprofit sector–the Population Council and others–do have research programs and some of their results will find their way into the marketplace. There have been advances, but with the exception of RU 486, which is a totally new concept, everything we have today is only an improvement on what we had in 1970. We have better oral contraceptives and better ways of delivering the hormones by injection and implant, plus much better IUDs, but we don't have anything new–the sponge notwithstanding–that is truly dramatic. If you look at any other field of medicine, what we used in 1970 and what we use today for almost any condition you can think of, is dramatically different. Not true in contraception.

Mailman School of Public Health, [722 West 168th Street, New York, NY 10032-2603, tel: (212) 305-3929, fax: (212) 305-1460, www.mailman.hs.columbia.edu].

Christopher Flavin, Worldwatch Institute

As president of Worldwatch Institute, Christopher Flavin is the institute's chief executive officer, representing the organization before a wide range of international audiences. In his long career at Worldwatch, he has helped guide the institute's development, and is actively engaged in international climate change and energy policy discussions.

Population and Consumption

World population has grown from 1.5 billion a century ago to 6.4 billion today. We've also increased our consumption level per person dramatically, so in effect the demand on the earth's resources has probably risen eight- to ten-fold over the last century. Population growth rates are falling in many parts of the world, yet we may well add another three billion people over the next century taking world population to roughly 9.5 billion people, causing additional stress on the planet's re sources. Then developing countries are increasing their consumption levels, whether through the purchase of appliances, automobiles or the adoption of high-protein diets heavy in meat similar to ours, which increases the amount of water and land resources required to meet those needs. As a result we may be looking at unsustainable demands on the earth's ecological systems over the next century.

If we grow to 11 or 12 billion, which is possible if fertility rates do not continue to decline, we are going to be facing miserable conditions in developing countries likely to experience most of that additional population growth, particularly India, parts of Africa and regions throughout the Middle East. This will result in declining standards of living, enormous healthcare problems and troublesome issues relating to international migration and terrorism generated by the tension caused by an increase in population density. In sum, the population pressures of the future will unleash global effects and not be confined to countries where most of the population growth occurs.

Obviously, countries become over-populated long before they actually fill every square meter with people. We require resources – water, wood and all variety of energy needs in the United States – which we are taking away from much of the rest of the world in order to support a high standard of living for fewer than 300 million people. Arguably, the world is already over-populated for it would take several planet earths to support the American standard of living for 6.4 billion people. Thus it seems our goal should be not only to stabilize population but over time to reduce the world population.

America's Role in Global Population Issues

In the U.S. we need to recognize that we have a population problem. We are growing much faster than most industrialized countries and still have relatively high fertility rates and significant immigration. Since each American uses vastly more resources than people in other countries, additions to the U.S. population cause far more problems in term of global resources than do increased populations in countries like India and China. Also, the U.S. population is now actually growing faster than is China's. Moreover, here in the United States, we should have much stronger foreign aide programs than we do. It's an embarrassment the U.S. is not stronger at supporting family planning, particularly educating young girls, providing healthcare and the other kind of social services that are a key ingredient to swollen population growth.

More important than slowing population growth in the United States is a reduction in our consumption levels. We use an enormous amount of resources per person compared to almost any other part of the world and if we are going to play a responsible role in terms of our leadership in any global sense, we need to significantly reduce our material resource consumption. We can do that without damaging the quality of American lives by behaving with greater care in terms of the impact that our own lifestyle, our own consumption pattern, even our own family size has on the earth's resources. Americans should be doing a lot more to help developing countries meet basic human needs, family-planning programs and the poverty reduction which is really key to developing a sustainable world.

U.S. levels of support for family and other social programs internationally have declined dramatically under the Bush administration. Such trends do not bode well for a more stable world. Growing population pressures and the demand placed on natural systems by humans are the main factors driving what is already one of the greatest massive extinctions the earth has ever seen.

Where We Go From Here

The world has made remarkable progress in slowing the rate of population growth over the last several decades in regions like Latin America and parts of Asia. Many industrialized countries in Europe have achieved demo graphic stability, some are even witnessing declining populations. High fertility nations in sub-Sahara Africa, the Middle East and South Asia are still experiencing numerous difficulties. We now need to help those challenged nations find ways to replicate what has already been accomplished in other parts of the world by providing a basic social safety net, as well as education, particularly for girls. This plus providing access to family-planning services can help stabilize the world's population during the next century.

Stabilizing population is an essential ingredient to creating a sustainable world, as is reducing consumption levels. We also need a new energy system based on renewable sources and hydrogen fuel and we need to develop more efficient ways of using water resources. If all these come together, a sustainable world is indeed possible. A key to sustainability is to think about global progress in new ways – not measuring our forward movement in terms of the size of the GNP, but rather to focus on the wellbeing of people, on the quality of their lives and the health of ecosystems. If we build an economy based more solidly on local resources, we are going to be better off and we will have a sustainable world for future generations.

Worldwatch Institute, [1776 Massachusetts Ave., N.W., Washington, D.C. 20036-1904, USA, tel: (202) 452-1999, fax: (202) 296-7365, e-mail: worldwatch@worldwatch.org].

Thomas Lovejoy, **Heinz Center for Science, Economics & Environment**

Founded in December 1995 in honor of the late senator, the H. John Heinz III Center for Science, Economics and the Environment is a nonprofit institution dedicated to improving the scientific and economic foundation for environmental policy by bringing together business, government, academia and environmental groups to collectively address environmental issues. Program areas include Environmental Reporting, Global Change, and Sustainable Oceans and Coasts. Thomas E. Lovejoy has been president of the center since May 2002. Before this he was at the World Bank and the United Nations Foundation. At the World Wildlife Fund U.S. he originated the concept of debt-for-nature swaps, and is the founder of the public television series *Nature.*

Biodiversity and Climate

The U.S. went to Rio not having signed the biodiversity convention. Ultimately it was signed in the Clinton administration, but never ratified by Congress, so that's something we are lagging behind on. Of course we aren't doing a lot under the climate convention either. Currently international thinking about the environment and sustainable development agenda has got a little muddy and we're losing sight of the importance of the goals of those two conventions. One is trying to stem the sixth great extinction in the history of life on earth of which you can already see the first stages. Another is doing something serious about climate change, which needs to be dealt with now, not 50 years from now.

When you hear a figure for the average temperature of the earth and the change in that average temperature, it sounds like a small number, but in terms of the climate system, it translates into huge effects. Four degrees is the difference between a mile of ice over New York City or the way it is today. Most of the projections of climate change say we're going way beyond four degrees in the warming direction unless we come to grips with things very soon. My real worry is that when the time comes for our species to move in response to climate change, they're going to be dealing with a landscape we've already turned into an obstacle course.

It seems the first recorded extinction as a consequence of climate change is a lovely golden toad from a cloud forest in Costa Rica where because of climate change, the cloud level which nurtures cloud forests has risen higher in altitude beyond that which the toad was accustomed to, so they couldn't get the moisture they needed and subsequently disappeared. There's a lot of grim news you see every day in headlines about deforestation and fires. In 1997 there were so many wildfires in various parts of the world, a smoke cloud as big as Brazil hung over South America.

Progress Despite the Bad News

What those headlines never seem to show is the progress being made on the other side of the ledger–some beyond anybody's wildest dreams. When I first showed up in the Brazilian Amazon, there was one national forest, one road and two million people. Today there are numerous roads and huge deforestation rates with 20 million people living in the Amazon, but at the same time a real effort in conservation at both the national and state levels has been moving forward. Recently the Brazilian government agreed to an ambitious project in coöperation with the World Bank and the World Wildlife Fund that when completed will put more than 40% of the Brazilian Amazon under some form of protection. Nobody would ever have predicted that. Still, it's not enough and it will be a race to the finish, but it proves that progress is possible. One can actually hope for, and obtain, results.

When you reduce your family size to two children, you get wealthy enough to buy a sports utility vehicle so you can now pump out the CO_2. Sustainable development involves serious issues of consumption and lifestyle. Basically, the world can neither support everybody living an American lifestyle nor living as hunter-gatherers. The answer lies in between. It's very complicated and those of us from the highly consumptive societies are going to have to find ways to do with less.

A crucial issue hinges on our current concept of what is a quality of life. It doesn't have to translate into any particular amount of fossil fuel consumption per day or any consumerism that exceeds our basic human needs in terms of water and other resources. We have to become much more creative in terms of finding ways to substitute our current high-natural-resource-consumption lifestyle patterns with those methods that are equally satisfying to human beings. These are not problems for the economy, they are opportunities.

Just look at the two halves of the island of Hispaniola–Haiti on one end and the Dominican Republic on the other. In Haiti you find an ecosystem which has been degraded to almost the lowest level possible with a huge amount of soil erosion, immense biodiversity loss and serious water-supply problems. It makes no sense to run a system down to the point where it's no longer supporting the community–but it's all about population, because Haiti's population is many times larger than the Dominican Republic's, which still has a fair amount of its forests and natural resources. All countries have problems, but these two nations, side by side, show you what population pressure is all about.

Epiphany

A great moment of truth in my life came in the 1980s when I realized if I wanted to make the maximum possible difference on behalf of life on earth, I couldn't do it being a field biologist; I had to deal with political and economic systems because that's where much of the driving forces affecting biodiversity actually stem from. I still need to go back to nature periodically, just a couple days will do it, to refill my spiritual batteries as it were so I don't get totally frustrated and beaten down by the day-to-day efforts.

H. John Heinz III Center for Science, Economics and the Environment, [1001 Pennsylvania Ave. NW, Suite 735 So., Washington, DC 20004, tel: (202) 737-6307, fax: (202) 737-6410, www.heinzctr.org].

Dr. Lester Brown, **The Earth Policy Institute**

The Earth Policy Institute was founded to raise public awareness to support an effective public response to the threats posed by continuing population growth, rising CO_2 emissions, the loss of plant and animal species and the many other trends that are adversely affecting the Earth. The dissemination of information from the institute is designed to help set the public agenda and its purpose is to provide a vision of what an environmentally sustainable economy will look like, how to get there from here, plus make ongoing assessments of this effort.

Lester Brown, president of the institute, began his environmental career in rural India where he became intimately familiar with the food/population issue. In 1959 he joined the U.S. Department of Agriculture's Foreign Agricultural Service as an international agricultural analyst and served as an adviser to Secretary of Agriculture Orville Freeman on foreign agricultural policy and administrator of the department's international agricultural development service. In early 1969 he established the Overseas Development Council and five years later founded the Worldwatch Institute, the first research institute devoted to the analysis of global environmental issues. In May 2001, he founded the Earth Policy Institute to provide a vision and a road map for achieving an environmentally sustainable economy. He has authored or coauthored 49 books which have been published in more than 40 languages and been recognized around the world with numerous prizes and awards for his efforts on behalf of the world's environment.

Population, Water and the Global Ecological Deficits

When *Building a Sustainable Society* was published in 1981, I was hoping we would get the brakes on population growth during the last two decades of the century so by 2020 we could be approaching population stability at six billion. I still think that was a great idea, probably even more so now than I did then. At the time I wrote this book, we had not even linked population growth with the water issue, for example.

Now we know water scarcity is emerging as one of the most serious resource issues facing the world today which we've never had to deal with before–aquifer depletion. Irrigation problems go back 6,000 years, but until 50 years ago most of the water pumping was done by animal-driven or human-powered devices. With the advent of the diesel pump and powerful electrically driven pumps, we now have the capacity literally to suck aquifers dry –and that's happening in many places around the world at essentially the same time. Water tables are dropping in China, India, the Middle East, North Africa, U.S. and Mexico where we're over pumping in order to satisfy current needs for food, because 70% of all the water we pump from underground or divert from rivers is used for irrigation, 20% is used by industry and 10% for residential purposes.

Over-pumping is a measure we use to increase current food production that almost guarantees a drop in future food production when that aquifer is depleted. We're beginning to see this now play out in Northern China where water tables are falling, including under the North China Plain that accounts for a quarter or more of China's grain harvest. Water shortages are now reducing China's grain harvest. From 1950 China's grain harvest went from 90 million tons to 392 million tons in 1999–an extraordinary growth. Then it dropped. In the last three years it has been under 350 million tons–a loss that's equal to the grain exports of Canada and Australia combined, so it's not insignificant. China has been offsetting this decline by drawing down its stocks, but it probably can't do that for more than another year or two, then it's going to have to turn to the world market.

When China goes this route, we will know it. Not necessarily because we're reading about it in the newspapers, but we'll see it at the supermarket checkout counter because when China comes to the world market for large quantities of grain, it will drive up the price of grain to levels we've not seen before. Rising grain prices may be the first economic indicator to signal serious trouble in the relationship between us, now six billion plus in numbers, and the earth's ecosystem, the natural systems and resources on which the economy depends.

In a country like Brazil with its rainforest, which is called that because it in effect produces rain with all the evaporation and transpiration from the vegetation, when it rains down on a healthy stand of rainforest, about 25% runs off into a stream and eventually goes back to the ocean while 75% evaporates and is carried further inland and comes down again. When you clear the rainforest for farming or grazing cattle, when the rain comes down, 75% runs off and goes back to the sea while only 25% comes back as evaporation to be recycled inland. So the rainforest is actually a huge conduit that recycles water inland. If we disrupt that process, then the western part of the Amazon will begin to dry up because rainfall will decline – which is starting to happen now.

There's a lot of marginal land being farmed in Brazil that should not be farmed because there's not enough rainfall to hold the soil, so it erodes, or it's land that's beginning to degrade because of the way its being farmed. Brazil is now in the process of clearing vast new areas of land on the southern and western edges of the Amazon Basin, an area known as the Cerrado which is a Savannah-type area with mostly grass but some small trees and shrubs. The pressures of Brazil's population, which are substantial – around 160 million and growing by a couple million a year – is beginning to affect the resource base in Brazil. We think of Brazil as being a large, sparsely populated country, but the amount of fertile productive cropland there is rather limited.

A number of things are happening in Brazil. It's expanding its production of soybeans at an extraordinary rate. This year for the first time the area in soybeans in cultivation will exceed the area in grain. The soybeans are being produced for internal use – as soybean meal for livestock and poultry production – but also for export. Together Brazil and Argentina produce as many soybeans as the United States – which also has more land in soybeans than in wheat. Soybeans are becoming a huge crop in the Western Hemisphere – and most of those soybeans are going to China, which is where the soybean came from originally.

China wants soybeans not only because of their burgeoning population, but because of the rising incomes the Chinese are moving up the food chain, consuming more pork and poultry, eggs and now even milk and dairy products. They require a lot of protein meal to feed these livestock and poultry, so interesting links are now evolving between major countries like Brazil, China and the United States as the effects of rising affluence puts pressures on world agricultural production patterns in the form of extraordinary demand for soybeans and grain.

One of my questions in looking at the world and the ecological deficits we're running up on so many fronts – whether it's overfishing, over-plowing, over-grazing, over-cutting or over-pumping – is that these deficits are beginning to converge in China creating a huge dust bowl in the northwestern and northern parts of the country with an expansion of their deserts. I read an embassy report entitled, "Mergers and Acquisitions," which one would expect to be about businesses, but it's about deserts in China that are expanding so much they're starting to merge. So we're seeing desertification and the formation of a dust bowl on a scale not seen before.

The principal reason for this is over-grazing. To put this in perspective, the U.S. has 97 million head of cattle. China has 127 million. We have 8 million sheep and goats. China has 290 million that are de-vegetating the country. The government is trying to plant trees, but not anything near the effort needed to turn this situation around. The Gobi Desert has expanded by 20,000 square miles in five years – that's about half the state of Pennsylvania. The desert is moving toward Beijing and is only 150 miles from the capital now giving the leadership in Beijing much concern.

We're seeing ecological meltdowns and environmental refugees on a scale never seen before in the world. Huge dust storms forming each year make life miserable in the cities of eastern China, including Beijing, and also in Seoul and in Japanese cities because this dust is carried eastward by the prevailing winds. In this country, during the dust bowl period, nearly three million people left the southern Great Plains, many going to California – the so-called Okie's of Steinbeck's *The Grapes of Wrath*. The same phenomenon is occurring in China. A U.S. Embassy official in Beijing who accompanied a relief shipment to Inner Mongolia and was expecting to see grassland, reported when he got there he found it was largely desert with huge numbers of refugees because the drifting sands eventually cover things up and people have to leave.

There are environmental refugees on many fronts now. Many villages in northeastern Iran have been evacuated because the water tables are falling forcing people to leave – basically water refugees. Each year the drop in the water table is greater than the year before because the rising demand curve for water looks like the population growth curve, except it's steeper. Over the last half century world population has doubled, but the demand for water has tripled. Once that demand curve climbs above the sustainable yield of the aquifer, then each year the gap between the demand for water and the sustainable yield gets larger. The drop in the first year is small, the next year it grows and each year it gets exponentially bigger – but the problem is underground and invisible.

What I think may happen someday with world grain stocks

now at the lowest level in perhaps 20 years, is we may wake up to discover there's not enough grain to go around and not enough water to produce enough grain anymore. In my opinion water is going to emerge as a major issue in the years ahead. With the globalization of the world economy, water scarcity now crosses national boundaries via the international grain trade.

The fastest growing grain market in the world in recent years has been North Africa and the Middle East: Morocco, Algeria, Tunisia, Libya, Egypt and the Middle East through Iran. Every country in that region is faced with water shortages and they're pressing against their limits. So when the demand for water in cities or industries escalates, those needs are met by taking water from irrigation, and then grain is imported to offset the loss of productive capacity. It takes 1,000 tons of water to produce one ton of grain, so the most efficient way to import water is in the form of grain. Last year the water required to produce the grain imported into North Africa and the Middle East was roughly equal to the annual flow of the Nile River, so if you visualize the water deficit in the region, it's like another Nile River flowing into the region in the form of imported grain.

With a country like China facing acute water shortages and the need to import large quantities of grain, there are now 1.3 billion people needing water subsidies. There may begin to unfold a dramatic relationship between China and the United States based on the politics of scarcity, because if China turns to the world market for a substantial share of its grain supply, it will necessarily turn to the U.S. which controls half the world's grain exports. This could drive up grain prices here because China, with a trade surplus with us of over 80 billion dollars a year, has enormous purchasing power here. They could buy our entire grain harvest if they wanted to. If 1.3 billion Chinese begin competing with us for our own grain, there will be political pressures here to restrict exports of grain. But how do we do that given China's importance in the world, its economic role and its military position? The problems relating to population growth of decades past are now beginning to manifest themselves in ways we never imagined when I was writing *Building a Sustainable Society* in 1981.

Climate Change

It's been claimed that future wars in the Middle East will more likely be fought over water than oil–but it's hard to win a water war. Competition for water is more likely to take place in world grain markets and it'll be countries that are financially the strongest, not those that are militarily the strongest, that will prevail in this competition. There are a number of other scenarios where populations are simply overrunning natural systems at the local level–forests, rangelands, croplands, fisheries or aquifers–where we're running up huge ecological deficits.

At some point we'll have to deal with these. We don't have good computer models that put these things together–underground water tables, population growth, land availability and climate. The world's farmers are now facing two new challenges no generation of farmers ever faced–aquifer depletion and a rise in temperatures. There are now a number of scientists, agricultural scientists, crop ecologists, plant breeders and agronomists working on the effect of rising temperature on crop yields. There've been a number of efforts to use climate models to try and project the effect on future agricultural production, but the modelers have been handicapped because we have no good data on the effect of higher temperatures on grain yields. We're now getting that. We're also getting the higher temperatures.

This year will either be the warmest or the second warmest year on record. The 15 warmest years since record-keeping began in 1867 have all come since 1980. But the three warmest years on record have come in the last five years, so this trend is, if anything, gaining momentum. What crop scientists are now concerned about is the effect of this on grain yields. In the past when grain has been short and prices have gone up, farmers have responded strongly by pumping more water and putting on more fertilizer. Now they can't increase water pumping much because if they do, they'll just hasten the day when the aquifers are depleted.

Rising temperatures are an added conundrum. Crop scientists say a one degree Celsius rise in temperature reduces crop yields by 10% in rice, wheat or corn. We have failed to realize that the world agricultural system is finely attuned to a climate system that has been remarkably stable for the last 11,000 years, since agriculture began. Suddenly we've set in motion changes, increases in temperature not seen since civilization began, increases that dwarf anything in recent geological history, so the crop scientists are now quite worried. Work being done on this issue at the International Rice Research Institute in the Philippines and by the Agricultural Research Service in the U.S. indicate that rising temperatures are now a real threat to farmers. Farmers are trying to dig their way out of the deepest hole they've ever been. This challenge may bring the wake-up call. If I were to identify a single economic indicator most likely to signal serious trouble in the relationship between the six billion plus people on this planet and the natural systems on which we depend, it would be rising grain prices.

A Sustainable World

From Rachel Carson forward, we've been looking at the problems and figuring out what we're against. As a result of her book, *Silent Spring*, we lobbied against a lot of pesticides–with good reason. But it's not always clear what we're for. I think we need a vision of where we want to go, because if we don't have a shared vision, we're probably not going to get there.

The exciting thing is that almost everything we need to do to build an eco-economy is already being done somewhere. Denmark is getting 18% of its electricity from wind. South Korea has reforested its mountains and hills in a remarkable way. Germany gets 72% of its paper from recycled fiber. The United States gets over half of its steel from scrap. The Netherlands has a very sophisticated urban-transport system that emphasizes the bicycle and public transportation where the automobile is third in a remarkable system that's working very well. Around the world everything that needs to be done is already being done by at least one country. Already 45 countries have stabilized their population. It's a matter of putting it all together in each country in order to build a sustainable economy. It's entirely doable. We just have to do it.

The time has come to pull out the stops. Shortly the Kyoto Protocol is going to be history because the events on the climate front are going to make it clear we've got to get serious about stabilizing climate and accelerating the transition from a carbon-based economy with fossil fuels to a hydrogen-based economy. Also quite doable. We have the technologies now, but we must start thinking about cutting carbon emissions by half over the next decade. Something of that order of magnitude is needed in order to stabilize climates sooner rather than later. Two areas of technological progress setting the stage for the restructuring of the world energy economy are the advances in wind turbine design that have reduced the costs from 38 cents a kilowatt hour 15 years ago to 4 cents or less today and the evolution of fuel cell engines. Once you get cheap electricity, you can electrolyze water to produce hydrogen – which is the fuel of choice for fuel cell engines. We've not quite grasped it yet, but the potential is clearly there.

Ecological Wake-Up Calls

Within the next few years, we'll be getting some serious wake-up calls on the climate front with storms more destructive than anything we've ever seen, grain shortages driven by water shortages and rising temperatures, ice melting plus a multitude of other events which could unfold quickly in the next few years. These events will get our attention. In the past, December 7, 1941, was a wake-up call. The day before, had you taken a public opinion survey asking whether the U.S. should get involved in the European war, the public would have been overwhelmingly against it. The next day this all changed. In his State of the Union Address, President Roosevelt laid out an ambitious war production plan—which we accomplished within two years—transforming the automobile industry into producing tanks and aircraft engines. It was extraordinary. No one could have projected that we could have restructured the economy almost overnight.

If we get a series of wake-up calls that ratchet up the level of concern, people eventually will demand we do something and things will begin to happen. Social change is not a linear process. Take for example Eastern Europe. In 1989 the Berlin Wall came down and there was a political revolution in Eastern Europe. You can search the political science journals during the 1980s, but no political scientist predicted something was about to happen. Not even the CIA. After it happened, no one could explain it. A social threshold was crossed and suddenly it became clear to everyone that the great socialistic experiment was finished. The one-party political system, the centrally planned economy, and even the people in power realized it, which is why there was essentially a bloodless political revolution, Romania being the minor exception. That change came very quickly and we could be moving towards something like that on the environmental and population fronts. Suddenly it's going to become clear to everyone that we have to do something. I spend almost all my time trying to track and understand these issues. It seems we're getting close to the point where the wake-up call is about to be heard.

Energy Futures

What we're talking about with energy is actually rather trivial compared with the restructuring we accomplished for WWII. We can do what needs to be done. We have the technologies and don't have to wait for anything. The challenge now is to have something in place, a plan of how to respond when the time comes. A century of globalization will be followed by a century of localization. Wind is abundant. Everyone has sunlight and most countries have geothermal energy. Only a handful of countries have most of the world's oil so we're looking at a much more equitable distribution of energy resources than we've had in a long time.

Solar cells are now becoming competitive with kerosene in India and with candles in Andean villages. There are roughly a million homes in the world getting their electricity from solar cells – almost all of them in developing countries. This number is going to grow exponentially so that the two billion people in the world who do not now have access to electricity will probably get their electricity from solar cells. As we turn to wind, solar cells and sunlight for energy, we'll begin to understand how intimately connected our future is with that of the earth. This will be a healthy development and lead to a profound shift, back to an earlier set of values where we understood how dependent we were on the earth and its natural systems and resources.

Earth Policy Institute, [1350 Connecticut Ave. NW, Washington DC 20036, tel: (202) 496-9290, fax: (202) 496-9325, e-mail: epi@earth-policy.org].

James Nations, Conservation International

Conservation International (CI) works to protect established areas of biodiversity, develop corridors between existing areas of biodiversity and protect endangered species. It also works within communities to meet human needs and preserve the environment. Their rapid assessment program employs a full contingent of biologists, tropical ecologists and specialists who visit ecosystems and exit with basic surveys of the ecosystem's condition as well as the species living within it. Their tropical ecology assessment and monitoring program is designed to create a series of permanent biological stations in different countries to accumulate standardized data via established protocols to document changes to these areas over the next five decades. Through their center for environmental leadership and business, CI partners with corporations helping them decrease their footprint on the environment via their marketing and products. James Nations, vice president of development agency relations for Conservation International, is developing big block funding for CI field programs from development banks and governments interested in conserving biodiversity.

Conservation and Population: A Case Study in Chiapas

The U.S. Agency for International Development has environmental specialists and population and reproductive health specialists. What we've done over the last four or five months is work with USAID to put a project together that pulls those departments together to create a population-environment project that is being put into place in three countries: the Philippines, Madagascar and Guatemala. In the case of Madagascar, we're setting up a population-environment program working with families that are situated in and around protected areas.

Mexico is another case. I went to Chiapas in the mid-1970s to work with a lowland rainforest Indian known as the Lacondon. I walked in an anthropologist and walked out three years later a conservationist because of what I learned during that time living with very traditional people. These are Maya Indians with long black hair, both men and women, who wear white cotton tunics and move through the rainforest barefoot. What I came to realize is that these beautiful people lived inside the rainforest and, in some ways, were the rainforest. Everything about them—their clothing, their houses, the instruments they used, their technologies—came from the indigenous plants. In learning about their agriculture, their lifestyles, I also realized they don't do what many people do in the tropics, which is replace the rainforest with something else in order to survive, instead, they've become part of the ecosystem themselves.

One night I had my epiphany, my transformation, sitting around a campfire with the family I was staying with. The head of the household was speaking candidly. "Look," he said, "we've been here for 1,000 years. As you've seen, we haven't killed the animals and we haven't destroyed the trees. The animals are still there, we guard them, and we guard the forest. Now we look up and we see people moving in from the highlands, following the logging roads. The loggers come in and take the mahogany and cedar trees. We see bulldozers coming in and making roads. We see cattle ranchers coming in and clear-cutting the rainforest and planting grass and raising cows and we don't know what to do about it." The gentleman looked me right in the eye and said, "Can you help us figure out what to do?" It's taken me 25 years to actually come up with a real response to that troubling, overwhelming question.

One of the projects we engendered here at Conservation International is an initiative to help the Lacondon Maya declare their territory a wildlife reserve so they can live inside it and have legal protection to keep loggers, cattle ranchers and colonists out. They can legally patrol the boundaries of what is now the only intact forest left in the region and keep that ecosystem alive and well, because they are part of the forest.

Another thing we're doing in Chiapas is looking at the whole social movement. Why are people moving down from the highlands into the territory of the Lacondon Maya? Much of it has to do with the degradation of lands in the highlands as well as the skewed land tenure in other areas of the state. It also has to do with the rapid rate of growth of the human population in that state. Women in Chiapas give birth and raise seven children on average. If, as is the case today, most of those children go on to be part of families that are slash-and-burn farmers, that means you have an increasing number of people out clearing tropical forests so that they can farm where the tropical forest used to be. That has a heavy impact on traditional people like the Lacondon Maya who are just trying to keep alive their culture, their language and the environment they've always been part of.

What we're ultimately about is keeping alive the biological foundations of life on earth. In Chiapas we work specifically on a protected area clustered around a 3,000 square-kilometer block of lowland rainforest, a biosphere reserve, the last large block of rainforest in Mexico. We work with the Mexican government and local nonprofit organizations based in Mexico to train guards to set up ecotourism locations in and around that protected area, to demonstrate the economic value that biosphere reserve has to the country of Mexico and to the people who live around there. Simultaneously, we know that people are pushing up against the edges of the biosphere reserve.

People don't invade protected areas because they're malicious or because it's fun to cut down trees. They're doing what you or I would do. They're trying to feed their kids; they're trying to keep their family educated and healthy. So, how do you counter a situation like that? It's not by erecting walls and it's certainly not by saying these people are evil. In fact, it's by working with those people by asking them what is it that we can do, how can we work together so that your needs are met, yet we can keep alive the natural resources that all of us depend upon, because this is where your food and your livelihood come from. This is our source of biological diversity, a wellspring of water, a clearinghouse for the very air we breathe. These are basic components of the very foundation that keeps us alive as a species on this planet.

We end up working on projects like ecotourism, setting up lodges that bring tourists in, conduits for income generation. Lodges get built with partly local staff. The people make money by protecting biodiversity rather than eradicating it. People will always tell you that they want to be assured of good health and of healthcare and that includes reproductive health. If you ask women, it turns out that they don't want to have seven children. They'd prefer to have two or three and be confident about their children's health and happiness. We can help them achieve that basic right by giving them access to the same kinds of healthcare, including reproductive healthcare, that women everywhere in the world should have. We're not trying to invent anything new.

How can we work with those communities so that we can maintain connectivity between the islands of biological diversity and the need to meet human requirements? We can work, for example, in the creation of vanilla or cocoa plantations. We've had great success with shade-grown biodiversity-friendly coffee. We're looking for ways you use the benefits of biodiversity to benefit local people, simultaneously improving their lives and protecting the biological foundation of life on earth.

We're also interested in helping indigenous people keep alive their homelands and territories so they can maintain their culture and traditions–and also the biological diversity they live within and amongst. We want to protect the millions of species we share the earth with–with protected areas, corridors and species. How you go about achieving those outcomes is where the reality sets in, because the most effective way to do that is by improving people's lives so that they see the benefits which come to them from achieving such outcomes. They need to understand how a national park can benefit local communities. Concomitantly, we must concentrate on ecological economics. If we can figure out how to work with communities so they can generate income by helping to protect the biological foundation of life, then we're on track.

One of the most valuable commodities traditional people have is their traditional knowledge. People have been living inside their natural ecosystems for thousands of years and understand the nuances of how ecology throughout those regions works. You can fly the best trained, most experienced tropical ecologist into a rainforest in Bolivia, Indonesia or Mexico and the first thing they'll do is find a local guide who can tell them what's really going on inside that forest. The scientist may know the Latin names of the species the guide points out, but the guide's going to lead them to the precise species and tell them something new about it: when it flowers, what animal eats it, what it's really good for in a practical sense. The most valuable commodity traditional people have is this kind of ethno-botanical knowledge. If we can help them take that information and experience and turn it into a sustainable commodity generating income for themselves and their community, then we've got something that works.

If Conservation International hadn't developed the campaign to save the "hotspots," we would have probably lost a number of incredibly diverse ecosystems in a range of countries throughout the world–that includes national parks and other protected areas. We would have witnessed the loss of cultural traditions of indigenous people who are very much allied with natural ecosystems. It takes thousands of dedicated people in key high-biodiversity-value countries confronting these challenges and environmental threats that we're facing in the 21st century in order to shape and devise strategies that can ensure the durability of indigenous traditions and the very lives of those people within precious ecosystems. How to protect the biological foundation of what is the only planet we have to live on is core to this vision.

When a surgeon injects you with curarine, a byproduct of curare, it paralyzes the heart and the lungs while you're put on a heart-lung machine so that the surgeon can get in and repair damage to your heart while it is motionless. When that naturally derived ingredient wears off, the surgeon and team removes you from the heart-lung machine and (hopefully) everything starts working again. That's a direct benefit of traditional knowledge from the jungles of a place like Chiapas.

Another example is the rosy periwinkle of Madagascar which turns out to be a major element in curing childhood leukemia. Another is the yew tree of the Pacific Northwest which has been showing great promise in ovarian cancer cures. The list goes on, to the point that some 25% of the products sold in the pharmacy across the street here in Washington DC actually originated in plants in the tropical forest. A quarter of those ingredients in drugs found in prescriptions originated in the tropical rainforests of the earth. Who wouldn't want to keep those forests alive?

Conservation International, [1919 M Street, NW, Ste. 600, Washington, DC 20036, tel: (202) 912-1000, www.conservation.org].

Shirley Hufstedler, *U.S. Commission on Immigration Reform*

Shirley Hufstedler became chair of the United States Commission on Immigration Reform after a long and distinguished legal career which included serving as special legal consultant to the attorney general of California in the complex Colorado River litigation before the U.S. Supreme Court. After serving as a judge in several venues, she was appointed by President Jimmy Carter as secretary of education before serving as chair of the immigration reform commission.

Critical Immigration Priorities

I think it is necessary to draw some lines on chain immigration whether it is a popular decision or not. A major priority of immigration reform should be reuniting husband, wife and children, and if the health issues are not a barrier, the grandparents. Adding cousins to that list is carrying it too far. People's relatives are important, but our country should not be required to absorb and care for more human beings than the infrastructure can bear. We cannot disadvantage native-born and naturalized citizens by giving greater priorities in funding and care to those who are not in the same situation.

Other factors need to be considered. Newcomers consume more than they did in their countries of origin because they had less available to them there. You need to ask what it does for those supplying these new consumers? What are they contributing to the country? It's a complex equation, involving many factors and you must figure out how to reduce the minuses to equalize some of the pluses, so you minimize the projected downside.

Population is something with which the Americans have never had to grapple with. Since the end of the so-called baby boom, we have lacked a needed domestic-labor force, so the country needs a documented immigrant population to fulfill that need. But examining the future, no one can predict accurately what it will be like 100 years from now. Yet my advice to young people is not to mourn about the current state of affairs, but rather to make up their minds and decide to do something about the problems.

Any of these issues can be debated and moved forward by getting together with other concerned folk who are doing things. There's not a campus in the country that doesn't have interest groups concerned with the preservation of the environment or interest groups dealing with political issues. There is not a religious group that has not committed time to dealing with many of these issues. The more you get involved, the less time you'll have to fret about the problems. There are things that are awful, but how can you feel good about yourself if you don't do something positive to make a difference? It can be a small contribution that can have significant results–so go to it. The children, after all, are the future of every country in the world. So if we think about what it's going to be like for them, we can address a whole array of issues and make an enormous difference in how the world will be.

Shirley M. Hufstedler, Senior of Counsel, Morrison & Foerester LLP, 555 West Fifth Street, Suite 3500, Los Angeles, CA 90013, tel: (213) 892-5804, fax: (213) 892-5454, www.mofo.com

Michael S. Teitelbaum, *Alfred P. Sloan Foundation*

The Alfred P. Sloan Foundation was established in 1934 by Alfred Pritchard Sloan, Jr., who was president and CEO of General Motors. It funds programs in science and technology, standard of living and economic performance, education and careers in science and technology, bioterrorism, federal statistics and civic programs.

Michael S. Teitelbaum, program officer for the foundation, is a demographer who is responsible for research fellowship pro grams in various fields. He served on the U.S. Commission on Immigration Reform, USCIR, when it concluded that a substantial flow of immigrations to the United States is in the nation's self-interest if that flow is well ordered and reforms are established to regulate it. The commission recommended additional visas be provided for immediate family of U.S. citizens and legal permanent residents to clear backlogged cases, while at the same cutting back on the number of visas provided to more remote family members.

The U.S. Commission on Immigration Reform

The USCIR was established by law in 1990 with a mandate to advise congress and the administration on necessary and desirable reforms in U.S. policy. The commission concluded that if the national interest is to be served by a legal immigration flow, it was very important irregular and unauthorized migration flows be brought under control.

A lot has taken place since September 11, 2001 when of the alleged 19 hijackers of planes that day, three had student visas but the majority did not. All had some kind of temporary admission visa, which indicates the temporary admission system was incapable of screening out people with such intent. There has been a strong effort by the federal government to rectify that situation but the actual success of that endeavor is unknown. By demonstrating the easy availability of fraudulent documents, the commission underscored a situation that makes it difficult to get control over unauthorized immigration.

The commission felt the broad contours of the legal immigration system made sense where priorities are given to family, needed skills and humanitarian admissions, but within those categories there seemed to be imbalances that should be rectified. In particular, there was concern about the immediate families of U.S. citizens and legal permanent residents–green card holders–who could not gain visas because of backlogs in the legal immigration system, while at the same time there were those with more remote family ties admitted because of the way the preference system works. The commission recommended additional visas be provided these family members so as to rectify those problems.

The United States is the largest country of destination for legal and illegal immigration, although Canada has a higher ratio of immigration to its population base. The subject of immigration brings out a passionate debate on all sides, yet after listening to all points of view expressed in open hearings, consultations and other venues, I concluded there is a remarkable absence of serious objective analysis of these issues.

A Conversation of the Deaf

Classical economists like Adam Smith and Thomas Malthus believed the core resource land was by definition limited in its supply. You can't create additional land and, therefore, if population continued to grow, it would outstrip the resources of the earth. In retrospect, that was rather simple-minded. They couldn't have known in 1800 that land could become enormously more productive or that new land could be discovered or irrigated. The industrial revolution had not even occurred.

By the 1860s, Karl Marx and his associates were writing about the supposed unlimited capacity of labor and socialist systems to produce goods, with the idea that there was no resource limit on human population as long as you had the right political and economic systems in place. In their view, the reason there was poverty, which they denounced vigorously, was because capitalists were siphoning off profits from the toil of the working class who were thereby impoverished. If you did away with the capitalists, workers would produce enough for their own wellbeing. This conversation of the deaf has been going on ever since and continues to the present day.

Among political conservatives in the U.S., you hear views very similar to traditional Marxism: there is no limit to re sources as long as you have the right political and economic system in place. Get these right and there's no limit to the number of people who can be supported. Then you have on the biological and ecological side of the argument those who disagree saying the earth is not infinite. You can destroy the environment and there are natural ecological and biological limits which we don't understand. This argument says there's no limit to the number of people as long as you have the economic system right. What about the ecological system? These two viewpoints do not communicate well with each other. It's another conversation of the deaf in the modern age.

I would suspect at the beginning of the 20th century no one could imagine in another 100 years the earth could support a 6.4 billion population. Yet the system, imperfectly and with a great deal of poverty, starvation and suffering around the world, is supporting this number. That's why it's so difficult to look forward another 100 years and have any real clear picture of what the world will look like.

The glass is always half empty and half full and it's easy to point to the wonderful things that have happened or the terrible things that have happened without recognizing the opposite. It's incumbent on us to be objective, measured and balanced on these issues, particularly since the answers are really unknowable. We can do our best and be as honest and objective as possible in making our projections, but we really have no way of knowing for sure what the outcomes will be. We're just trying to model things as best as we are able and we're not all that able.

[Alfred P. Sloan Foundation, 630 Fifth Avenue, Suite 2550, New York, NY 10111, tel: (212) 649-1649, fax: (212) 757-5117, www.sloan.org.]

Gloria Feldt, *Planned Parenthood Federation of America*

The Planned Parenthood Federation of America (PPFA) is both an advocacy organization and a provider of healthcare services. There are 126 PPFA affiliates managing 875 health centers around the country. It has the largest grassroots activist and donor base of any advocacy group focusing on reproductive rights with millions of active donors and serves nearly five million women, men and youth who receive reproductive health and/or education services from it each year. Planned Parenthood works in the international service arena through its international division. Planned Parenthood Global Partners links U.S. affiliates with family-planning programs abroad for the purpose of shared learning interchange.

Gloria Feldt became the president of PPFA in 1996. A mother of three children by the age of 20, Ms. Feldt graduated from college in 12 years and became a Head Start teacher and activist in the civil rights and women's movements. She has served the organization for almost 30 years, since starting in 1974 in West Texas. At PPFA she developed voter-education programs and established a federal PAC to support pro-choice candidates.

Planned Parenthood

In the beginning Planned Parenthood was completely funded by USAID funds. Then in the Reagan and the first Bush eras we chose not to accept USAID funds because of the accompanying Gag Rules and were one of the few organizations that decided we could not compromise our principles in that way. As a result, we lost 30 million dollars a year, yet we gained a smaller, more agile program that has nobody telling us what to do.

We are looking to find out where the real needs are, particularly what other organizations won't do or where others won't or can't go, but we can. Our global partners match our affiliates in the U.S. with family-planning programs in other countries for the purpose of shared learning interchange. Sometimes we are able to bring project resources to the developing world, but it's the interchange that's most important.

Since Roe v. Wade

Planned Parenthood is rapidly becoming the largest provider of abortion services in the country, if it isn't already. One reason for this is that physicians have been so under attack. When they are part of Planned Parenthood, they are a part of a community that provides them a base of support that keeps them going–spiritually as well as practically speaking. This success has enabled some of our affiliates to begin various training programs of their own and many of our affiliates across the country are now providing training for physicians who didn't receive it in medical school but who want to be able to provide this important reproductive healthcare service to women.

Planned Parenthood is also engaged in advocacy, working closely with medical students for choice all over the country. The best way to bring young physicians into the movement is by giving young medical students a chance to carry out their commitment to women's healthcare and reproductive healthcare. We thus foster support systems where these young physicians can come, often on a residency rotation, to see how family-planning services, in their broadest sense, are provided–not just abortion.

Most medical schools don't do a very good job of teaching physicians how to deal with family planning and sexuality issues. They don't teach doctors how to counsel people about the different birth control methods to help them decide what's best for them. They even don't teach them how to be comfortable talking with patients about their sexuality. We try to fill all of those roles. Our name is well-known and we provide a space where people gravitate. It's a combination of providing a place where people can get high quality abortion services while working with medical schools and physicians' groups to make sure doctors are always coming into the movement. It is also about providing a safe space from a public policy realm for people to be able to support the services.

Sexuality and Contraceptive Choice

We are a long way from having a society where sexuality is viewed as it is in Western Europe–as a healthy, normal part of human life. That's not the way our culture is. It seems that parents of young children and teens today, who mostly grew up in the 1960s and 1970s, want to do a better job, They know they didn't get the kind of sex education they needed and they want to be able to talk to their kids about sex, so they seek help and organizations like ours are here to help them. There is also a great deal of information available through the internet. We have our website: www.teenwire.com. You don't need school board approval to go to this website, nor do you have to jump through political hoops to use the web.

A great deal of the reduction in teen pregnancy has come from long-acting birth control. Injectables have made a tremendous difference. Teens who use family-planning clinics are likely to choose long-term contraception and the injectables. It's easier for them in many ways because they might forget to take a pill or have a condom with them. So it's very effective. This is why I believe Planned Parenthood's network of almost 900 health centers across the country provides an opportunity to expand that geographic accessibility and it's just phenomenal. Now most of our surgical abortion-providing affiliates are offering Mifepristone, or early medical abortion. Our hope is to enable all of our medical centers to be able to do so.

Emergency contraception is another reason why our teen pregnancy rate is declining. This technique has been known for over 20 years, but only recently has the FDA given its imprimatur so we can talk about it publicly. If every woman of reproductive age had access to emergency contraception at all times, we could reduce the unintended pregnancy rate and the abortion rate in the United States by one half. The health, economic and emotional consequences of this common technology could make such a difference. Emergency contraception should be available on every street corner. That is why we at Planned Parenthood have tried to make it easily accessible for people to get. We even suggest women get a prescription for emergency contraception before they may need it, as a precaution. This, too, could help bring down the rate of unintended pregnancy, teen pregnancy and abortions.

There's a big challenge today because we have a generation of people who've grown up not knowing what life was like before "Roe v. Wade." Telling someone women can die is very different from those of us who actually experienced knowing a friend or seeing someone die as a result of an illegal abortion. We face the special challenge of getting the message across of how important it is for women to have this fundamental human and civil right to make their own childbearing choices without government intervention. Right now in the United States, the threats to basic reproductive freedom are intense, not just to abortion services.

The legal principles underlying "Roe v. Wade" are exactly the same legal principles underlying "Griswold v. Connecticut" that gave us the right to birth control. So the most basic decisions–how many children you're going to have and how you are going to space them and your bodily integrity, which is the most basic human right any of us have–are all at dire risk. For the first time the presidency, both houses of congress and many of our lower federal courts are aligned and ready to take away this basic right to choose. George W. Bush has had two opportunities to appoint justices to the U.S. Supreme Court. No one knows how the balance of the court will be tilted when these new judges cast their votes. We could be facing a sea change in our rights with the demise of reproductive freedom as we have known it in this country.

Education and the Right to Choose

We must advance an aggressive agenda to get sex education to our children–and not just the abstinence-only variety congress is currently funding with 103-million dollars a year. They need real, honest, comprehensive sexuality education and we need to make sure our laws and policies affirm, protect and support the things we thought "Roe" was going to do–which now are at risk. We must codify that basic principle giving us the right to determine how many children we have and when we have them–not the U.S. Congress, nor the state government, nor the supreme court. We as individuals have the human right to decide.

Almost every day I get a letter or hear a story how having access to reproductive healthcare has improved someone's life. "You saved my life," is the phrase I often hear. One letter that I have carried with me for years was from a 16-year old who had an unintended pregnancy. After she chose to have an abortion, she wrote, "You enable people's dreams to come true. I have learned how making one little mistake can change your whole life. When I have a child, I want to be able to give it all the things that a child should have, and I want to thank you because you have given me a chance to live and to make my life what it should be and the life of my future children what it should be." That is the wind beneath my sails.

As I have traveled around the world, I have seen that women everywhere want the same thing–a better life for their children and they will do almost anything to get this for them. They also want a better life for themselves. So I have seen that with reproductive healthcare services, if you build it they will come. I have yet to see someplace where services are available in a community that is culturally sensitive where women haven't found a way to get there and use those services, somehow. They then go tell their friends. There's this kind of silent revolution that goes on.

If we're going to have the kind of world we want our children to inherit, we had better step up our work on reproductive healthcare and family planning. There are still over a 120-million couples in this world who would like to be able to use contraception but simply don't have access to it because the resources aren't there. Probably there are many more than that who have some level of difficulty getting to those services. It's so inexpensive; it does so much good for so little. It's such a great humanitarian gift to provide and it's something we can do and know how to do. We deserve to have a world where everyone who wants access to family planning and reproductive healthcare services can get it. We have the ability to do that. For the sake of our future generations, we must do this.

[Planned Parenthood Federation of America, 434 West 33rd Street, New York, NY 10001, tel: (212) 541-7800, fax: (212) 245-1845, www.plannedparenthood.org.]

Suzanne Olds, *Marie Stopes International*

Marie Stopes International is an organization registered in Great Britain as a trust charity which collaborates with partners in 38 countries, most in the developing world. MSI provides a full range of reproductive health services, including safe abortion. It's goal and mission is the motto, "Every Child a Wanted Child." Peripheral services include well-baby services, obstetric care services, HIV/AIDS services and family-planning services. Marie Stopes' clinics are known worldwide by their blue doors, the quality of their services and friendliness of their staff. Suzanne Olds is the

U.S. representative for MSI. Her career in reproductive health and family planning included posts at USAID and Pathfinder International.

Nine Billion Condoms

Nine billion condoms will be required by the year 2015 just to accommodate family-planning needs. That doesn't even include the situation with HIV/AIDS. There's simply no way these billions of contraceptives can be funded at this time, at least according to the current systems, and every single donor knows it. There is going to be an incredibly large shortfall of all sorts of reproductive health supplies within the next ten years.

However, if you take the array of population- and reproductive-health organizations collectively, compute their contributions, and divide that by every couples' average expenditures for annual birth control, the numbers look different: we find the costs to be quite low. Moreover, donor organizations and clinics may recover certain costs. We charge every client. Although some are subsidized very heavily, some of our clinics are totally self-sustaining and make enough money to pay their operating costs.

In sub-Saharan Africa we have an excellent program in 19 of the 23 districts spanning Malawi, for example. This is a nationwide program where we are providing services primarily with paraprofessionals. We don't have to have doctors and these trained paraprofessionals help keep the costs down, plus it allows us to go into rural areas where doctors just aren't going. We have an equally successful program that recently started in South Africa where we now have six clinics with nine more slated to open with the help of a generous U.S. benefactor. Unfortunately, in Kenya, we were receiving USAID money and because of our position on the provision of safe abortion, this USAID money was cut and we had to terminate services, which is terribly problematic for Kenya.

Overpopulation is a major deterrent to security and to the

economic viability of many of the countries that we work with in sub-Saharan Africa and Asia. It doesn't take an Einstein to figure out that overpopulation is a big threat in many countries where their economic opportunities are lacking. This is the moment that we all have to be concerned about this.

It seems as though the Bush Administration is bound and determined to strip women of their reproductive rights, not just here in this country, but worldwide. We see it in what appears to be very insidious ways like their abstinence-only programs. Abstinence only for young women and men in sub-Saharan Africa is wishful thinking? I agree that we need to teach abstinence, but that's a death warrant for many young people which we can lay at the door of the Bush Administration.

Hope

There are reasons to be hopeful. Look at Bangladesh where in 1982 there was hardly a woman employed and little family planning. Today there are many women who are employed, especially in the textile industry where they are earning decent wages and so now don't have to provide a dowry to their husbands (just the opposite). The number of children these women are having has decreased from about six down to around three. They're planning their families and there is hope.

There are any number of countries where women are beginning to enjoy more rights, well on their way to full rights and full education for literacy. The hope is that we're getting smarter; we're learning from past success stories and utilizing those data to make better and better decisions.

[Marie Stopes International, 153-157 Cleveland St., London W1P 5PG, England, tel: (44 20) 7574 7400, fax: (44 20) 7574 7417, Suzanne Olds: soldsmd@aol.com, www.mariestopes.org.uk.]

Linda Martin, Population Council

John D. Rockefeller 3rd founded the Population Council in 1952 to address international population issues and specialize in three areas of research: biomedical, health services and social science. Although first and foremost a research organization, the Population Council also works to build the capacity of individuals and institutions around the world in providing technical assistance and has 18 offices around the world. Recent developments in the biomedical division include research of male contraceptives, vaginal rings bearing hormones, and microbicides that would be prevent the sexual transmission of HIV.

Now 30% of the council's work is focused on HIV/AIDS, applying our research to improve service delivery in areas ranging from prevention, voluntary counseling, testing, peer education and programs to reduce mother to child transmission. It is also used as a tool to provide treatment in resource-poor settings and addressing the issue of AIDS orphans in Africa. Biomedical research is exploring prevention of transmission between heterosexual partners via microbicides. Linda G. Martin, PhD, previous president of the Population Council, came to the council from the nonprofit policy research organization, RAND.

Young Women's Lives

Of all the people on earth today, roughly a billion of them are between 15 and 24. It's the largest cohort ever of adolescents. The council has worked to broaden the perspective in the field and try to understand the lives of these young people in a broad socioeconomic context and then develop interventions that will help make the most of the resources that are available for them. The world's future depends on how good a start these young people get in their lives. We are very proud of the fact that of the four long-lasting reversible contraceptive methods on the market today, we developed three of them. The Copper T-IUD is really the workhorse of those contraceptives.

In addition, the council is interested in other spheres of a young person's life as well – sports activities that give young women, in particular, a greater sense of confidence about themselves, their bodies and the choices they make in the future. We're also doing vital work with women in the workforce as the garment industry has moved around the world. It is a major employer of young ladies and we've been interested in places like Bangladesh studying female garment workers to understand what their work experience means for their lives. The choices women make surrounding marriage and whether they even have a choice about marriage and their future family development is also a deep concern to us at the council.

Helping People Achieve Their Goals

One of projects that really is making a difference is the work we've been doing in Ghana. The Navrongo Project there has been so successful the government is now scaling it up over the entire country. It seems you could not ask for any better outcome from a policy research perspective than having your ideas taken and applied throughout a nation.

Africa is an area where we're strongly engaged in trying to improve access and quality of reproductive healthcare. The statistics on total fertility rates and survey data on unmet needs makes us aware there is still a large proportion of couples in Africa who are not able to achieve their family size goals. So that's a priority for us. Of course, national averages can be deceiving. Nor can we forget that besides Africa, there are many people in other areas of the world like Asia and Latin America living in poverty, without access to good maternal and child healthcare services. The Population Council certainly has a role to play in those countries in trying to improve the services that the poorest people receive and to help them achieve their own goals for their families.

John Bongaarts, *Policy Research Division, The Population Council*

Since John Bongaarts came to the Population Council in 1973, his research has focused on a variety of population issues, including the determinants of fertility, population-environment relationships, the demographic impact of the AIDS epidemic and population policy options in the developing world.

Three Divisions

The Population Council has three divisions–the biomedical division, which develops contraceptives; the second works with governments overseas on operational research and the third division is the policy research division which does social science and demographic research in order to understand the determiners of demographic processes, particularly those of rapid population growth, and the consequences of demographic trends. Once we have an understanding, we try to provide insight into the programs and policies that might be most effective in dealing with adverse consequences of population growth.

In the 1950s, rapid population growth became our first concern. The United Nations made its first global projections and expected huge increases in the world's population, particularly in the developing world. Some argued for coercive measures, others felt incentives would work while still others said we could do it voluntarily through family planning. The basis on which this last stance was taken came from research done in the 1960s and 1970s by the Population Council.

We found when we interviewed women in developing countries asking them about their reproductive lives, their preferences, how large a family they wanted and whether or not they were using contraception, that a large portion of women wanted to stop or postpone childbearing, but they were not using contraception. That phenomenon was called a KAP (Knowledge, Attitude and Practices) gap. It's now called the unmet need for contraception.

The obvious thing to do was to help this substantial group of women who didn't want to get pregnant get access to contraception and give them information about it. This research led to a huge increase in funding for family-planning programs for those governments in the developing world wanting to pursue that approach. This worked so well that by the late 1970s and during the next decade, fertility in the developing world has declined dramatically to some half as many children per family.

Those observers who believe we've won this battle because fertility has declined by such a large amount assume population growth is now close to zero. That turns out to be incorrect, because even though fertility has declined greatly, we need to go to two children per family. The last step in the fertility decline is difficult to achieve. As long as women have more than two children, population growth continues. Besides this, health conditions are improving around the world. The longer women live, the larger the population size will be. Also, we cannot forget that there is a momentum to population growth due to a young population age structure. In many developing countries, half the population is now under age 18.

Global warming is the result of excess production of greenhouse gasses–and there is a straightforward relationship with population growth. The more people produce those greenhouse gasses which accumulate in the atmosphere, they in turn contribute to the warming of the planet. The projections are now for a significant increase in warming over the next century that is likely to have serious consequences. One is the rise in the sea level which will impact those living near the coast. Bangladesh and Egypt are examples of countries with large concentrations of people near the coastline. Another effect will be on agriculture for in the long run higher temperatures in the grove areas will lead to diminishing yields that will make it even more difficult to raise enough food to feed growing populations. Global warming is going to be the topic of conversation 50 to 100 years from now.

Population growth has a wide range of environmental effects. Air pollution has gotten very serious and is causing major health effects. The shortage of water means the competition for water between industry and agriculture is growing very rapidly especially in the most densely populated areas of the world – South Asia and the Middle East. Then population pressures acerbate the depletion of natural resources, deforestation and the loss of biodiversity. Rapid population growth also contributes to slow economic growth and poverty in various ways – most obviously being when you have a large number of people competing for a limited number of jobs, wages are kept low and this contributes to poverty.

No one can predict what will happen when the nine billion people which the United Nations now projects will live on our planet in 2050. Some revisions to this number are being made, but they depend on how rapidly fertility declines in the developing countries – all unknown factors. Fertility in the developing world has declined rapidly in the last two decades but to project that on developing nations for the next two or three decades seems to be unduly optimistic. There are several countries where fertility has now stopped going down like Bangladesh and Egypt where fertility leveled off a bit above three children per woman. That's a trend that could spread to other countries which would make the United Nations' projections be on the low side.

To bring fertility down in countries where fertility is three or above requires a reduction in unwanted childbearing plus a reduction in wanted childbearing. Unwanted childbearing is happening still in about one in five births. The most obvious way to address this is to assist couples to implement their preferences by giving them access to family-planning methods. Even if you could eliminate all unwanted fertility, which is virtually impossible, you still have high levels of wanted fertility. A key factor in bringing this down is increasing levels of education. There are large differences between women with no schooling and women with primary and secondary levels of schooling in terms of the desired family size. In places like Bangladesh where a large proportion of women have no schooling, the desired family size is likely to remain above two unless the level of schooling changes. So we must both strengthen family-planning programs and emphasize schooling.

The onset of childbearing and long-range fertility is becoming a key issue because in many developing countries where fertility has declined, the remaining cause of population growth is now momentum. This is a new phenomenon we are uncertain how to deal with. So far our research suggests the key way to reduce momentum is to delay childbearing, which in turn means delaying the first marriage. We're now researching the factors which determine the timing of marriage. In virtually every society there's a strong correlation between a girl's level of schooling and her age at marriage. If she stays in school through secondary or higher education, she often marries in her mid-twenties. A girl without schooling often marries in her mid-teens–which carries enormous implications for the future welfare of the girl and her family and to societal issues because of the demographic implications of momentum.

The HIV epidemic now has become extremely serious in parts of south and east Africa. Fully a third of the adult population is infected. With no access to effective treatment, these individuals will all die with an accompanying massive increase in mortality. Some have concluded that these populations will be wiped out or will go into steep declines. That turns out to be not the case, at least not in the majority of these populations. While there will be doubling or tripling of the deathrates, the birthrates in most of these countries is still higher than a triple deathrate. As long as the birthrate is higher than the deathrate, there is population growth.

Some thought overpopulation and shortage of resources would doom civilization as we now know it. In the third millennium, generally speaking, standards of living are rising, life expectancy is rising, so the future does not seem quite as bleak as the pessimists had expected. On the other hand, environmental degradation is rising everywhere and environmental problems are becoming more severe. In the long run, those most densely populated areas of the world, South Asia and the Middle East, are bound to face serious environmental, economic and societal problems.

[Population Council, One Dag Hammarskjold Plaza, New York, New York 10017 USA, tel: (212) 339-0500, fax: (212) 755-6052, e-mail: pubinfo@popcouncil.org, www.popcouncil.org.]

Robert Engelman, *Population Action International*

Population Action International (PAI) is a privately funded nonprofit focused on research on population issues and advocacy with governments to develop policies that result in human development and slowing population growth worldwide. PAI connects population to environment issues by encouraging legislators to enact population policies that improve the environment and also by joining environmental groups with local community development organizations to help couples have fewer children so they can concentrate on their livelihood in order to give their families an improved quality of life.

Robert Engelman, vice-president for research at PAI, emphasizes analyzing the links among population dynamics, population policy, reproductive health, the environment and human development. He has written widely on population's connections to renewable fresh water, cropland, fisheries, forests, climate change, biodiversity, economic growth and community development. He chairs the board of the Center for a New American Dream, a nonprofit organization dedicated to enhancing the quality of North American life while reducing consumption of natural resources.

Reproductive Health and the Environment

There are many projects around the world in villages and developing countries helping people improve their crops, plant trees and preserve clean water for their children's sake. As men have migrated to the big cities to find jobs, women increasingly are stepping forward and are doing their farming. What they're saying to the various organizations helping them is that though they're learning to do all kinds of new things to improve their livelihood, they can't really put them into effect unless they're able to prevent the next pregnancy.

Since we're in the middle of a revolutionary demographic transition around the world, even in the lowest income developing countries women want to have fewer children than before, certainly fewer than their mothers had and fewer even in many cases than their older sisters had. Based on our research on these relationships, PAI is effectively trying to offer assistance to organizations interested in connecting reproductive health to natural resource management. We're trying to get the story out and make sure congress knows how important these issues are so that more international development assistance is actually focused on bringing the pieces together: better access to reproductive health with conservation of the natural resources that people depend on for their lives.

When Conservation International started working in Guatemala they received requests from the USAID to give the Guatemalan public officials training in reproductive health. The connections between these issues seemed obvious to all involved. In Nepal I visited a site where World Neighbors, a group out of Oklahoma City, was helping a community pipe water from the source of a stream on a mountain top down to their village but they were working with groups that had organized essentially to help coördinate family planning as a result of efforts that were made by the Family Planning Association of Nepal. Again, these people were addressing both issues at the same time. They saw their needs related holistically.

This approach to development is interesting more and more organizations in this work. Congress a year ago passed an appropriations bill directing the USAID to spend a portion of its money for family planning specifically in areas where there were endangered species. This is the first example I know where the U.S. linked population to environment at the level of congressional legislation signed by the president. So now USAID has taken an interest in these sorts of projects and will likely be funding work in Guatemala, perhaps Madagascar and maybe in the Philippines to further these sorts of connections.

The Poverty-Carbon Conundrum

Some countries such as Gabon, Syria, Turkey and Cuba are actually somewhat sustainable in terms of their emissions of carbon dioxide, so they're not contributing in a major way to the warming of the planet. However, it's because of their poverty not their philosophy. This is an ironic connection we need to break – making them less poor and the wealthier countries more sustainable. But how to do that? Those in wealthy countries need to change things about their technology and the way they live. One possibility might be to tax carbon and apply revenues from this to produce sustainable technologies that can be shared with the lower-income countries of the world that need an infusion of equity and economic help for their development – which needs to be developed in sustainable manners. An important part of the sustainability dilemma is approaching replacement fertility.

Migration

Migration is one of the three main components of population change – besides births and deaths. In looking at world population on a global basis, migration is irrelevant because you can neither arrive nor leave the earth. From a national perspective it makes a big difference, and it is one of the major population dynamics right now in the developing world. At the local level, there's often tension between refugees and their host populations over concerns about environmental degradation. In Africa, for example, wildlife preserves with gorillas and chimpanzees are threatened by refugees fleeing from a conflict.

Whether environmental degradation has contributed to international migration is another question. Do people emigrate to flee a degraded environment or are their reasons based on issues such as family reunification or better economic prospects? Certainly, some environmental factors are involved in the equation and then once migrants arrive in the developed nations, their average consumption is likely to rise as compared to that of the country they came from.

The population of the United States is growing at one of the fastest rates in the industrialized world – which is at about the same rate the world population is growing, 1.2%. Part of this is population momentum because even though the U.S. has roughly a replacement level fertility, there are a lot of young people having children right now. Even with no migration, the population would still be rising. The pressures on the future will be compounded by immigration factors. This all impinges on the environment and climate issues, given that everyone in the U.S. are all such heavy consumers.

Not Without Hope

It's obvious that population growth expands the scale of all human needs that end up interacting with the environment. To the extent that we achieve a peak to population growth worldwide, all of these factors will be eased. In fact, human population is one of the few positive megatrends going on globally. Women are having fewer children than ever before in human history. If that trend can be encouraged, supported and accelerated in the context of human rights and development, this will benefit all the other trends that are currently going in the wrong direction. This suggests that our situation is not without hope and there is some possibility we can turn things around in time. If there were only a greater sense of seriousness about these threats by policymakers and the administration, we could do more starting now. This would give at least a reasonable chance that our grandchildren could have a wonderful future. But we've got to act now.

[Population Action International, 1300 19th Street, NW, 2nd Floor, Washington, DC 20036, tel: (202) 557-3400, fax: (202) 728-4177, www.populationaction.org.]

Appendix

Dancing Star Foundation

The Dancing Star Foundation, founded in 1993, is a nonprofit public benefit corporation devoted to animal welfare, global biodiversity conservation and environmental education. Its primary emphasis is the maintenance of various sanctuaries which provide a refuge for wild habitat and species, as well as domestic animals. In addition, the foundation is involved in population and ecological research and analysis, book publications, lectures, colloquia and environmental filmmaking. For more information, contact:

www.dancingstarfoundation.org

Population Communication

Population Communication was founded in 1977 to determine the policies and programs needed to achieve population stabilization. It encourages world leaders to support the *Statement on Population Stabilization*, which has been signed by 75 heads of government. It also develops and tests cradle-to-grave small-family policies and programs in seven countries, concentrating on child survival, adolescent health and birth spacing. By contracting with a manufacturer in Taiwan, Population Communication donates maternal- and child-health medical equipment to 41 countries. Their feature-length documentary *No Vacancy* chronicles population policies, family planning and health programs in China, France, Italy, Ghana, India, Indonesia, Iran, Mexico, the Netherlands, Nigeria, Mexico and the U.S. *No Vacancy* may be ordered from:

Population Communication, 1250 East Walnut Street, Suite 220, Pasadena, CA 91106. (626) 793-4750, popcommla@aol.com.

GOVERNMENTS

Australia Agency for International Development (AusAID)

AusAID is an Australian government agency within the department of foreign affairs and trade which manages their official overseas aid program whose objective is to advance Australia's national interest by helping developing countries reduce poverty and achieve sustainable development. AusAID provides policy advice and support to the minister and parliamentary secretary on development issues, and plans and coördinates poverty reduction activities in partnership with developing countries.

www.ausaid.gov.au

Austria Ministry of Foreign Affairs Section for Development Coöperation

The government of Austria spends approximately 7.2 million euros for bilateral technical development coöperation each year. Projects aimed at a sustainable improvement of living conditions of the population in developing countries can be financed by a combination of funds coming from a private organizations and the foreign ministry. The Austrian department section for development coöperation within the foreign ministry provides assistance with proposals and offers advice during the project implementation phase.

www.bmaa.gv.at

Belgium Directorate-General for International Co-operation (DGIC)

The DGIC is responsible for policy preparation and evaluation. This aid administration has been working on strategic sectors in education, health, agriculture, food security, basic infrastructure, societal consolidation and conflict prevention, as well as in areas like gender equality, social economy and ecological sustainability.

www.belgium.gov.be

Canada **Canadian International Development Agency (CIDA)**

CIDA's mandate is to support sustainable development in developing countries in order to reduce poverty and contribute to a more secure, equitable and prosperous world. The agency's work is concentrated in the poorest countries of Africa, Asia and Latin America. It also supports democratic development and economic liberalization in Eastern Europe and the former Soviet Union and supports international efforts to reduce threats to international and Canadian security.

www.acid-cida.gc.ca

Denmark **Danish International Development Agency (DANIDA)**

Since 2001, the Danish government has worked to reorient its development assistance to reflect current global development challenges, threats and opportunities with a view to ensuring maximum impact and sustainability of the development initiatives. The promotion of sustainable development through poverty-oriented economic growth is the fundamental challenge for its development coöperation, yet its development policy is central and integral to its foreign policy.

Denmark's development assistance is focused on a select number of developing countries, working with efficient, long-term national strategies for poverty reduction and with a select number of multilateral organizations. Sub-Saharan Africa remains the main recipient of Danish aid. The government constantly works to ensure that Danish assistance fulfills its original goal of helping the poor by providing critical investments in education and health, infrastructure plus support for the development of a private sector as an engine for growth.

www.um.dk

European Community

The objective of the European Community is to foster sustainable development designed to eradicate poverty in developing countries and integrate them into the world economy by pursing policies that promote the consolidation of democracy, the rule of law, good governance and respect for human rights. Putting equity at the center of its policies, the directorate general for development gives priority to defending the interests of the most disadvantaged developing countries and the poorest sections of the population in economically more advanced developing countries.

www.europa.eu.int

Finland Department for International Development and Coöperation

The department of international development coöperation is part of the ministry of foreign affairs of Finland. The purpose of any funded project must be in accordance with the Finnish development policy and thus must increase the economic, social and mental well-being in the beneficiary country. Their aim is to empower the poorest groups while promoting social equality, democracy and human rights with activities that should also be environmentally sustainable.

www.global.finland.fi

France Ministry of Foreign Affairs

The Ministry of Foreign Affairs is responsible for all of France's international relations, including coöperation. Its main funding goes to nonrestricted aid.

www.france.diplomatie.fr/org

Germany Foreign Ministry

Germany's foreign ministry is responsible for all external and international relations of their government and explicitly covers their external and development policy with regard to the European Union, the U.N. and its aid programs. They are also responsible for the international representation of the German government. Grants for small-scale projects for technical coöperation are given out by embassies and consulates in developing countries.

www.auswaertiges-amt.de

Federal Ministry for Economic Coöperation and Development

Germany's federal ministry for economic coöperation and development handles development assistance with different funding lines. This ministry supports bilateral governmental coöperation and works with nongovernmental organizations making intragovernmental agreements for development assistance.

www.gtz.de

Japan International Coöperation Agency (JICA)

JICA implements bilateral aid programs and currently has population projects in Indonesia, Egypt, Turkey, Kenya, Thailand, Mexico, the Philippines and Peru. These projects include a broad range of activities from family planning service delivery to population education and maternal and child healthcare.

www.jica.go.jp

Organization for International Coöperation in Family Planning (JOICFP)

JOICFP conducts international coöperation programs in the fields of population, reproductive health, family planning and maternal and child health. It enjoys consultative status with the UN/ECOSOCand works closely with the Japanese government, JICA and other international agencies. JOICFP received a UN Population Award in 2001.

www.joicfp.or.jp

Luxembourg Ministry of Foreign Affairs

The ministry tries to foster the quality and quantity of NGO coöperation by working with officially registered organizations. Funding priorities include technical assistance, economic and industrial coöperation, environmental coöperation, regional coöperation, cultural and scientific coöperation, human rights and democracy, development education and social action including health, housing, education, vocational training and promotion of women's rights.

www.mae.lu

The Netherlands **Ministry of Foreign Affairs**

Investment in healthcare, water and social development represent the major part of the country program budget. Funding priorities focus on women, children and adolescents. Projects focused on women must contain elements relating to the platform for action from the closing document of the World Conference on Women in Beijing.

www.minbuza.nl

Dutch Organization for International Development Coöperation (NOVIB)

An honest income. Enough food. Clean drinking water. A proper education. Safety and freedom of speech. Millions of people around the world do not receive what they are entitled to. People are fighting for these rights in many countries and we join them in this by supporting local development projects, influencing the policy of national and international governments and organizations and by campaigning in the Netherlands. This approach enables people in developing countries to stand on their own two feet and is aimed at achieving a sustainable result. NOVIB works together closely with the eleven sister organizations of Oxfam International and with more than 3,000 local organizations. Together we form a worldwide movement of people with a single, communal goal: a just world which is free of poverty for everyone.

www.novib.nl

Norway **Norwegian Agency for Development Coöperation (NORAD)**

The Norwegian Agency for Development Coöperation is a directorate under the Norwegian Ministry of Foreign Affairs. Its most important task is to participate in the international coöperation to fight poverty. NORAD contributes to the effective management of development funds to ensure that the Norwegian development coöperation is of the highest quality.

www.norad.no

Sweden Swedish International Development Coöperation Agency (SIDA)

SIDA is a government agency that reports to the ministry for foreign affairs. Its goal is to improve the standard of living of poor people and, in the long term, to eradicate poverty.

www.sida.se

Switzerland Swiss Agency for Development and Coöperation (SDC)

The primary philosophy of SDC is to fight poverty through participatory programs, creating sustainable improvements in peoples' lives by involving them in the process. Its main intentions are to improve access to education and basic healthcare, to promote environmental health, to encourage economic and governmental autonomy and to improve equity in labor. Regions of work are primarily concentrated in 16 countries in Africa, Latin America, Eastern Europe, Middle East, and Asia.

www.sdc-gov.ch

United Kingdom Department for International Development, (DFID)

The DFID is the arm of the UK government that manages Britain's aid to poor countries and works to get rid of extreme poverty. It is headed by a cabinet minister, who is also one of the senior ministers in the government. DFID has two headquarters (in London and East Kilbride) and 25 offices overseas. Almost half of the 2,500 staff work abroad.

www.dfid.gov.uk

United States Agency for International Development (USAID)

USAID is an independent government agency that receives overall foreign policy guidance from the secretary of state. It supports long-term and equitable economic growth and advances U.S. foreign policy objectives by supporting agriculture and trade, global health and democracy, conflict prevention and humanitarian assistance. It provides assistance in four regions of the world: Sub-Saharan Africa; Asia and the Near East; Latin America and the Caribbean; Europe and Eurasia. With headquarters in Washington, D.C., USAID's strength is its field offices around the world where it works in close partnership with private voluntary organizations, indigenous organizations, universities, American businesses, international agencies, other governments and other U.S. government agencies. USAID has working relationships with more than 3,500 American companies and 300 U.S.-based voluntary organizations.

www.usaid.gov

FOUNDATIONS

For a more complete listing of foundations that fund population, reproductive health and family planning, go to the Funders Network, www.fundersnetwork.org .

Robert Sterling Clark Foundation

Since 1952 the Clark Foundation has developed program guidelines that have evolved and changed. Currently the foundation funds projects protecting reproductive rights, ensuring access to comprehensive reproductive health information and services.

www.rsclark.org

Compton Foundation

The Compton Foundation focuses most of its grant making in the areas of peace and security, environment and sustainability and population and reproductive health, with a special emphasis on projects that explore the interconnections between these categories.

www.comptonfoundation.org

Geraldine R. Dodge Foundation

The mission of the Geraldine R. Dodge Foundation is to support and encourage those educational, cultural, social and environmental values that contribute to making our society more humane and our world more livable.

www.grdodge.org

Ford Foundation

The Ford Foundation, headquartered in New York, funds projects worldwide and smart-growth related activities domestically. The foundation has programs in asset building and community development; peace and social justice; and education, media, arts and culture.

www.fordfound.org

General Service Foundation

The directors believe the foundation should concentrate on addressing basic long-term problems in the world: international peace, reproductive health rights and resources.

www.generalservice.org

George Gund Foundation

The foundation makes grants in the areas of the arts, civic affairs, economic development, education, environment and human services. Since 1995 it has evolved an increased interest in and emphasis on making grants to address urban sprawl.

www.gundfdn.org

William and Flora Hewlett Foundation

The foundation focuses its resources on activities in education, environment, global development, performing arts and population. In addition, it has programs making grants to advance the field of philanthropy and support disadvantaged communities in the San Francisco Bay Area.

www.hewlett.org

Robert Wood Johnson Foundation

The Robert Wood Johnson Foundation approaches the smart growth and community livability field from a perspective of public health.

www.rwjf.org

Henry J. Kaiser Family Foundation

The Henry J. Kaiser Family Foundation concentrates on major healthcare issues facing the nation and is a source of facts and analysis for policymakers, the media, the healthcare community and the general public.

www.kff.org

John D. and Catherine T. MacArthur Foundation

The MacArthur Foundation makes grants through two integrated programs – human and community development and global security and sustainability – plus the general program which undertakes special initiatives and supports projects that promote excellence and diversity in the media and the Mac Arthur Fellows Program which awards fellowships to exceptionally creative individuals regardless of field of endeavor.

www.macfound.org

Andrew W. Mellon Foundation

The Andrew W. Mellon Foundation currently makes grants in five core program areas: higher education, museums and art conservation, performing arts, conservation and the environment and public affairs.

www.mellon.org

Moriah Fund

The Moriah Fund seeks to promote human rights and democracy, help disadvantaged people gain self-sufficiency and control over their lives, foster sustainable development, promote women's rights and reproductive health and protect and preserve the environment.

www.moriahfund.org

Charles Stewart Mott Foundation

The Charles Stewart Mott Foundation, headquartered in Flint, Michigan, makes over $150 million in grants per year in the program areas of civil society, pathways out of poverty, environment and Flint-area grants, as well as exploratory and special projects. Its mission is "to support efforts that promote a just, equitable and sustainable society."

www.mott.org

David and Lucile Packard Foundation

The foundation provides grants to nonprofit organizations in the program areas of conservation, population, science, children, families and communities, arts and organizational effectiveness and philanthropy.

www.packfound.org

Pew Charitable Trusts

The Pew Charitable Trusts serves the public interest by providing information, policy solutions and support for civic life.

www.pewtrusts.com

Public Welfare Foundation

The Public Welfare Foundation is a non-governmental grant--making entity dedicated to supporting organizations that provide services to disadvantaged populations and work for lasting improvements in the delivery of services that meet basic human needs. Grants have been awarded in the areas of criminal justice, disadvantaged elderly and youth, environment, population, health, community and economic development, human rights and technology assistance.

www.publicwelfare.org

Rockefeller Foundation

The Rockefeller Foundation's grant making is organized around four program themes – creativity and culture, food security, health equity and working communities.

www.rockfound.org

Scherman Foundation

The main areas of interest of the Scherman Foundation are the environment, peace and security, reproductive rights and services, human rights and liberties, the arts and social welfare.

www.scherman.org

Wallace Global Fund

The Wallace Global Fund's mission is to promote an informed and engaged citizenry, fight injustice, protect diversity of nature and the natural systems upon which all life depends.

www.wgf.org

NON-GOVERNMENTAL ORGANIZATIONS

Advocates for Youth

Advocates for Youth works to increase the opportunities for and abilities of youth to make informed, responsible decisions about their reproductive and sexual health.

www.advocatesforyouth.org

AIDSinfo

AIDSinfo is a U.S. Department of Health and Human Services (DHHS) project providing information on HIV/AIDS clinical trials and treatment. It is the result of merging two DHHS projects – the AIDS Clinical Trials Information Service (ACTIS) and the HIV/AIDS Treatment Information Service (ATIS). AIDSinfo is a resource for current information on federally and privately funded clinical trials for AIDS patients and those infected with HIV. AIDS clinical trials evaluate experimental drugs and other therapies for adults and children at all stages of HIV infection – from those who are HIV-positive with no symptoms to those with various symptoms of AIDS.

www.aidsinfo.nih.gov

Alan Guttmacher Institute

AGI's ongoing national and international programs seek to

balance research, policy analysis and public education to enhance and defend reproductive rights of all women and men, with particular attention and concern for those who may be disadvantaged because of age, race, poverty, education or geographical location.

www.agi-usa.org

American Association of Sex Educators, Counselors and Therapists (AASECT)

AASECT is a not-for-profit, interdisciplinary professional organization. In addition to sexuality educators, sex counselors and sex therapists, AASECT members include physicians, nurses, social workers, psychologists, allied health professionals, clergy members, lawyers, sociologists, marriage and family counselors and therapists, family planning specialists and researchers, as well as students in relevant professional disciplines. These individuals share an interest in promoting understanding of human sexuality and healthy sexual behavior.

www.aasect.org

American College of Obstetricians and Gynecologists

ACOG works primarily in four areas: serving as a strong advocate for quality healthcare for women; maintaining the highest standards of clinical practice and continuing education for its members; promoting patient education and stimulating patient understanding of and involvement in medical care; and increasing awareness among its members and the public of the changing issues facing women's healthcare.

www.acog.org

American Public Health Association

APHA is the oldest and largest organization of public health professionals in the world, representing more than 50,000 members from over 50 occupations of public health.

www.apha.org

Association of Reproductive Health Professionals

ARHP members are professionals who provide reproductive health services and education, conduct reproductive health research and influence reproductive health policy. They include physicians, advanced practice clinicians (nurse practitioners, nurse midwives and physician assistants), researchers, educators, pharmacists and other professionals in reproductive health. The organization reaches this broad range of healthcare professionals both in the U.S. and abroad with education and information about reproductive health science, practice and policy.

www.arhp.org

Association for Women's Rights in Development

AWID is an international membership organization connecting, informing and mobilizing people and organizations committed to achieving gender equality, sustainable development and women's human rights. Their goal is to cause policy, institutional and individual change that will improve the lives of women and girls everywhere by facilitating ongoing debates on fundamental and provocative issues as well as by building the individual and organizational capacities of those working for women's empowerment and social justice.

www.awid.org

Californians for Population Stabilization

CAPS works to formulate and advance policies and programs designed to stabilize the population of California at a level which will preserve a good quality of life for all Californians.

www.capsweb.org

Carrying Capacity Network

CCN conducts an advocacy-oriented program focusing on solutions, such as achieving national revitalization, population stabilization, immigration reduction, economic sustainability and fiscal integrity and resource conservation.

www.carryingcapacity.org

Center for Development and Population Activities

CEDPA is a women-focused, nonprofit international organization founded to empower women at all levels of society to be full partners in development.

www.cedpa.org

Center for Environment and Population

The Center for Environment and Population is a nonprofit organization addressing the relationship between human population and environmental impacts. Through groundbreaking publications and a range of activities, the center strengthens the scientific basis of policies and public outreach to achieve a longterm sustainable balance between people and the environment.

www.cepnet.org

Center for Health and Gender Equity

CHANGE works to ensure that the population and health policies of international institutions supported by the United States government actively promote women's reproductive and sexual health.

www.genderhealth.org

Center for Immigration Studies

This center is the nation's only think tank devoted exclusively to research and policy analysis of the economic, social, demographic, fiscal and other impacts of immigration on the U.S. The center's mission is to expand the public's knowledge and understanding of the need for an immigration policy that gives first priority to the broad national interest. The center is animated by a pro-immigrant, low-immigration vision which seeks fewer immigrants but a warmer welcome for those admitted.

www.cis.org

Center for Reproductive Health Policy Research, University of San Francisco

The UCSF Center for Reproductive Health Research & Policy was formed in 1999 to address the health, social and economic consequences of sexual behavior by integrating research and training efforts in contraception and family planning with work that addresses sexually transmitted infections and HIV/AIDS. The center strives to develop preventive solutions to the most pressing domestic and international reproductive health problems.

www.reprohealth.ucsf.edu

Center for Reproductive Law and Policy

CRLP's domestic and international programs engage in litigation, research, policy analysis and public education in efforts to achieve women's equality in society and ensure that all women have access to appropriate and freely chosen reproductive health services.

www.crlp.org

Center for Research on Population and Security

Primary interests of the center are the study of the national and global security implications of overpopulation, the development of new and improved methods of contraception, evaluation of the safety and efficacy of contraceptive methods, the role of abortion in population growth control and the dissemination of research findings and related educational materials.

www.quinacrine.com

Centers for Disease Control and Prevention

The CDC is recognized as the lead federal agency for protecting the health and safety of people at home and abroad, providing credible information to enhance health decisions and promoting health through strong partnerships. It serves as the national focus for developing and applying disease prevention and control, environmental health and education activities designed to improve the health of the people of the U.S.

www.cdc.gov

Committee on Population, National Academy of Sciences

The National Academy of Sciences established the Committee on Population in 1983 to bring the knowledge and methods of the population sciences to bear on major issues of science and public policy.

www2.nas.edu/cpop

Conservation International

Conservation International (CI) utilizes an ever-growing scientific database from field research across the world's 35 biological hot spots to help find conservation solutions. Its Center for Applied Biodiversity Science, Center for Conservation and Government, various conservation funding methods and partners, Center for Environmental Leadership in Business and Population programs all contribute significantly to the protection of habitats, biodiversity and indigenous cultures and communities.

www.conservation.org

Contraceptive Research and Development Program

CONRAD is dedicated to improving reproductive health, particularly in developing countries where the need is greatest, by supporting the development of better, safer and more acceptable methods to prevent pregnancy and sexually transmitted infections (STIs), including HIV/AIDS. The program offers financial support and technical assistance for various stages of product development. Research is conducted at CONRAD's preclinical facility and Clinical Research Center at Eastern Virginia Medical School (EVMS) in collaboration with investigators at universities, research institutions and private companies worldwide.

www.conrad.org

Corporate Social Responsibility Newswire

CSR is defined as the integration of business operations and values whereby the interests of all stakeholders – investors, customers, employees and the environment – are reflected in the company's policies and actions. Corporate responsibility is no longer a luxury for companies. In today's global economy, it is critical for them to embrace social and environmental responsibility in order to meet the demands of their investors, consumers, employees and the communities they serve. CSRwire seeks to promote the growth of corporate responsibility and sustainability through solutions-based information and positive examples of corporate practices

www.csrwire.com

Development Associates

Founded in 1969, Development Associates has established capabilities to provide technical assistance, evaluation and program management in reproductive health, primary healthcare and health-care-management reform.

www.devassoc.com

DKT International

DKT International implements nine social marketing programs in eight countries: Brazil, China, India (two programs), Indonesia, Malaysia, Ethiopia, Vietnam and the Philippines.

www.dktintl.org

Emergency Contraception Website

The Emergency Contraception Website is operated by the office of population research at Princeton University and by the Association of Reproductive Health Professionals. This server is designed to provide accurate information about emergency contraception derived from medical literature.

www.not-2-late.com

EngenderHealth

EngenderHealth works worldwide to improve the lives of individuals by providing reproductive health technical assistance, training and information with a focus on practical solutions that improve services where resources are scarce.

www.avsc.org

Environmental Defense

Environmental Defense is dedicated to protecting the environmental rights of all people, including future generations. Among these rights are clean air, clean water, healthy food and flourishing ecosystems. It is guided by scientific evaluation of environmental problems and the solutions advocated are based on science. It works to create solutions that win lasting economic and social support that are nonpartisan, cost-effective and fair. Recognizing that low-income communities and communities of color have been disproportionately exposed to many environmental threats, it seeks answers that are equitable and just for all. As a U.S.-based organization, Environmental Defense pays special attention to America's environmental problems and their role in causing and solving global problems while aiming to share their findings internationally.

www.environmentaldefense.org

Family Care International

FCI helps shape international policies on women's reproductive health, produces and disseminates information tools.

www.familycareinternational.org

Family Health International

FHI is among the largest and most established nonprofit organizations active in international public health with a mission to improve lives worldwide through research, education and services in family health.

www.fhi.org

Family Planning International Assistance

FPIA, the international service division of the Planned Parenthood Federation of America, was established in 1971 to respond to the reproductive health needs of developing nations.

www.ppfa.org

Federation for American Immigration Reform

FAIR is a national, nonprofit, public-interest membership organization of concerned citizens who share a common belief that our nation's immigration policies must be reformed to serve the national interest. It seeks to improve border security, stop illegal immigration and promote immigration levels consistent with the national interest with more traditional rates of about 300,000 a year.

www.fairus.org

Futures Group

The Futures Group is a for-profit consulting firm dedicated to enhancing sustainable international development through the application of innovative policy, marketing, communications, education, training and research techniques.

www.tfgi.com

Global Health Council

The Global Health Council's mission is to promote and improve people's health worldwide through advocacy, by building global alliances, and by communicating experiences and best practices among those on the frontlines of global health.

www.globalhealthcouncil.org

GreenBiz.com

The nonprofit, nonpartisan GreenBiz.com works to use technology to bring environmental information, resources and tools to the mainstream business community. It aims is to provide clear, concise, accurate and balanced information, resources and learning opportunities to help companies integrate environmental responsibility into their operations in ways that combine ecological sustainability with profitable business practices.

www.GreenBiz.com

Ibis Reproductive Health

Ibis's goal is to produce stronger consumers of reproductive health products, services and policies, along with more responsive health personnel and systems. It aims to increase the reproductive health choices open to women and enhance their autonomy in exercising these choices. They conduct original research – both clinical and social science – analyzing and critiquing policies and protocols that limit reproductive choice and interpret and disseminate research findings that can help shape laws, policies and practices.

www.ibisreproductivehealth.org

Institute for Development Training

The mission of IDT is to improve the health of women, particularly in developing countries, by strengthening training systems especially with international ecumenical church networks.

www.nuteknet.com/idt

Institute for Reproductive Health

The Institute for Reproductive Health is dedicated to helping women and men make informed choices about family planning, providing them with simple and effective natural options. As part of Georgetown University's School of Medicine, the institute conducts research to develop natural methods of family planning and test them in service-delivery settings.

www.irh.org

International Council on Management of Population Programmes

ICOMP is dedicated to improving population program management. Its members include the heads of national and large NGO programs and management institutions. Founded in 1973, it operates through an extensive network of individuals and institutions from its secretariat based in Kuala Lumpur, Malaysia. For over 20 years it has assisted in sensitizing top managers, trained middle-level managers, created a network of management-related institutions in the field, promoted women's programs and encouraged South-South collaboration. It has also documented and disseminated state-of-the-art knowledge and experiences in the field of population program management.

www.icomp.org.my

International Fellowship on Population & Development of Rotary International

International Fellowship on Population & Development of Rotary International members share a common belief that humanity is now at a crisis point and that the interlocking issues of population pressure, environmental degradation and poverty must be addressed.

www.rispd.org

International Planned Parenthood Federation

IPPF is the world's largest voluntary family planning organization. With a membership of more than 140 national family-planning associations, it is active in promoting and supporting population and family planning activities in over 170 countries. It was founded in 1952 in Bombay, India, and incorporated by an act of Parliament in the United Kingdom in 1977.

www.ippf.org

International Population Program, Sierra Club

The International Population Program of the Sierra Club's main activities deal with public education, legislative campaigns promoting population stabilization and grass-roots organizing.

www.sierraclub.org

International Training & Health Program (INTRAH)

Over 24 years, IntraHealth has earned an international reputation for innovative performance improvement and training activities that benefit a wide variety of healthcare workers in such areas as family planning, reproductive health and safe motherhood. Much of its work is managed and implemented through its four international regional centers in Africa, Asia and Latin America and 14 country offices. Currently, the organization has active programs in 30 countries worldwide.

www.intrahealth.org

International Union for the Scientific Study of Population

IUSSP promotes scientific studies of demography and population-related issues. Originally founded in 1928 and reconstituted in 1947, it is the leading international professional association for individuals interested in population studies. Its network includes almost 2,000 members worldwide, one third of whom are from developing countries.

www.iussp.org

International Women's Health Coalition

IWHC works to ensure the sexual and reproductive rights and health of women of all ages by providing technical, financial, managerial and moral support to women's organizations, advocacy groups, health and rights activists and service providers in Africa, Asia and Latin America.

www.iwhc.org

Ipas

Ipas works globally to improve women's lives through a focus on reproductive health.

www.ipas.org

John Snow, Inc.

JSI is dedicated to improving the health of individuals and

communities in the United States and around the world by providing high-quality technical and managerial assistance to public health programs worldwide.

www.jsi.com

Johns Hopkins Center for Communication Programs

The Johns Hopkins Bloomberg School of Public Health established CCP in 1988 to consolidate health communication programs originating in the 1970s and early 1980s and focus attention on the central role of communication in health behavior change. With a staff of approximately 450 in the field and at its Baltimore headquarters, CCP has active programs in more than 30 countries worldwide as well as in Baltimore. It relies on a variety of partners to successfully implement its programs and activities. These partners range from small faith-based groups working in developing countries to multinational corporations. It appreciates the help of its many donors and partners worldwide, including private foundations, UN agencies, corporations, bilateral agencies, international and local NGOs and, most especially, the USAID.

www.jhuccp.org

Management Sciences for Health

MSH is a private, nonprofit education and scientific organization working to close the gap between what is known about public health problems and what is done to solve them.

www.msh.org/pop/pophome.html

National Abortion Federation

The National Abortion Federation is the professional association of abortion providers in the United States and Canada who believe women should be trusted to make private medical decisions in consultation with their healthcare providers. NAF currently offers quality training and services to abortion providers and unbiased information and referral services to women.

www.prochoice.org

National Association of Nurse Practitioners in Women's Health

NPWH's mission is to assure the provision of quality healthcare to women of all ages by nurse practitioners. It defines quality healthcare to be inclusive of an individual's physical, emotional and spiritual needs and recognizes and respects women as decision-makers for their healthcare. Its mission includes protecting and promoting a woman's right to make her own choices regarding her health within the context of her personal, religious, cultural and family beliefs.

www.NPWH.org

National Audubon Society

Audubon is a prominent environmental organization with a long tradition of conservation accomplishments in the United States. It maintains a strong programmatic focus on population, especially as it relates to the society's mission to conserve and restore natural ecosystems focusing on birds, other wildlife and their habitats for the benefit of humanity and the earth's biological diversity.

www.earthnet.net/~popnet

National Campaign to Prevent Teen Pregnancy

The mission of the National Campaign to Prevent Teen Pregnancy is to improve the wellbeing of children, youth and families by reducing teen pregnancy. High rates of teen pregnancy burden not only teenagers but also their children, families and communities while imposing large costs on taxpayers as well. To reduce teenage pregnancy, the campaign provides a national presence and leadership to raise awareness of the issue and attract new voices and resources to the cause. It provides concrete assistance to those already working in the field and also tries to ease the many disagreements that have plagued both national and local efforts to address this problem.

www.teenagepregnancy.org

National Family Planning and Reproductive Health Assoc.

NFPRHA has served as an important source of advocacy, education and training for the family planning and reproductive healthcare field for more than 30 years. Its mission is to assure access to voluntary, confidential, comprehensive, culturally sensitive family planning and reproductive healthcare services and to support reproductive freedom for all. It represents providers of care: public, private, domestic and international as well as researchers, educators, consumers and advocates. Its members provide reproductive healthcare services at nearly 4,000 clinics to more than 4 million women annually.

www.nfprha.org

National Resources Defense Council

NRDC is one of the nation's most effective environmental-action organizations which uses law, science and the support of more than a million members and online activists to protect the planet's wildlife and wild places and to ensure a safe and healthy environment for all living things.

www.nrdc.org

OBYGN.net

OBGYN.net has been designed for the specific needs of professionals interested in obstetrics and gynecology, the medical industry and women everywhere. Its goal is to improve services continuously to help in the delivery of Women's Healthcare. It offers the resources available at a medical conference or women's health symposium for at OBGYN.net one can learn about new techniques and innovations, investigate new opportunities, acquire published materials, network with peers, interact with medical professionals and women globally and shop for products and services.

www.OBGYN.net

Office of Population Research, Princeton University

The OPR at Princeton University is a leading demographic research and training center with a distinguished history of contributions in formal demography and the study of fertility change. In recent years there has been increasing research activity in the areas of health and wellbeing, social demography, and migration and urbanization.

www.opr.princeton.edu

Office on Women's Health

The OWH in the Department of Health and Human Services (HHS) is the government's champion and focal point for women's health issues which works to redress inequities in research, healthcare services and education that have historically placed the health of women at risk. It coördinates women's health efforts in HHS to eliminate disparities in health status and supports culturally sensitive educational programs that encourage women to take personal responsibility for their own health and wellness.

www.4woman.gov/owh

Pacific Institute for Women's Health

The Pacific Institute for Women's Health takes a comprehensive approach to the complex realities of women's lives and works through applied research, advocacy, community involvement, technical assistance and training.

www.piwh.org

PATH–Program for Appropriate Technology in Health

PATH creates sustainable, culturally relevant solutions enabling communities worldwide to break longstanding cycles of poor health by collaborating with diverse public- and private-sector partners to provide appropriate health technologies and vital strategies that change the way people think and act.

www.path.org

Pathfinder International

By providing access to quality family planning and reproductive health information and services, Pathfinder works to halt the spread of HIV/AIDS, to provide care to women suffering from the complications of unsafe abortion and advocates for sound reproductive health policies in the U.S. and abroad.

www.pathfind.org

Population Action International

PAI advocates the expansion of voluntary family planning, reproductive health services and educational and economic opportunities for girls and women.

www.popact.org

Population Association of America

The Population Association of America is a scientific and educational association of 3,000 professionals working in population studies and demographic research. Its focus is promoting scientific exchange and educating the public on demographic findings with annual meetings, publications and educational briefings.

www.popassoc.org

Population Communications International

The mission of PCI is to slow human population growth through entertainment-education programs which contribute to the improvement of social and environmental conditions worldwide.

www.population.org

Population Connection

Overpopulation threatens the quality of life for people everywhere. Population Connection is a national grassroots organization that educates young people and advocates progressive action to stabilize world population at a level that can be sustained by the earth's resources.

www.populationconnection.org

Population-Environment Balance

Established in 1973, Population-Environment Balance is a grassroots membership organization dedicated to public education about the adverse effects of continued population growth on the environment which advocates measures that would promote population stabilization in the United States.

www.balance.org

Population Institute

Founded in 1969, the Population Institute dedicates its efforts exclusively to creating awareness of international population issues among policymakers, the media and the public.

www.populationinstitute.org

Population Reference Bureau

For more than 70 years, PRB has been informing people about the population dimensions of important social, economic and political issues by providing timely and objective information on U.S. and international population trends and their implications.

www.prb.org

Population Resource Center

The Population Resource Center aims to further development of public policy by bringing the latest demographic data to policymakers through policy briefings and small-group discussions. Their programs help inform the debate and serve as a bridge between the social science community and the world of public policy.

www.prcdc.org

Population Services International

Population Services International uses social marketing to deliver health products, services and information that enable low-income and other vulnerable people to lead healthier lives in almost 70 countries.

www.psiwash.org

Program for Appropriate Technology in Health

PATH creates sustainable, culturally relevant solutions, enabling communities worldwide to break longstanding cycles of poor health. By collaborating with diverse public- and private-sector partners, they help provide appropriate health technologies and vital strategies that change the way people think and act to improve global health and wellbeing.

www.path.org

Rainforest Alliance

The mission of the Rainforest Alliance is to protect ecosystems and the people and wildlife that depend on them by transforming land-use practices, business practices and consumer behavior. Companies, coöperatives and landowners participate in programs to meet rigorous standards that conserve biodiversity and provide sustainable livelihoods.

www.rainforest-alliance.org

Reproductive Health Technologies Project

The mission of the RHTP is to advance the ability of every woman to achieve full reproductive freedom with access to the safest, most effective and preferred methods for controlling her fertility and protecting her health. RHTP views technology not as an end in itself, but as an essential means to giving women that control. Each emerging technology requires careful analysis of what is safe, effective, ethical, acceptable and appropriate with the understanding that the definitions of these terms and the implications of the technology vary from person to person and group to group.

www.rhtp.org

Sexuality Information and Education Council of the U.S.

SIECUS has served as the national voice for sexuality education, sexual health, and sexual rights for almost 40 years. It believes that sexuality is a natural and healthy part of life and that all people have the right to the information, skills and services they need to make responsible sexual decisions.

www.siecus.org

Social Contract Press

The Social Contract Press is an educational publishing organization advocating open discussion of such issues as population size and rate of growth, protection of the environment and precious resources, limits on immigration, as well as preservation and promotion of a shared American language and culture.

www.thesocialcontract.com

South-South

South-South collaboration is a worldwide strategy to improve reproductive health and family planning through sharing of experiences, successes and failures of program development of NGOs in developing countries. Collaborative efforts on transferring knowledge between developing countries are encouraging due to increased expertise in program implementation of developing countries and the similarities between existing conditions in these countries. With promising results, mechanisms to facilitate bilateral or multilateral sharing of experiences was developed and is referred to as the modalities of collaboration.

www.icomp.org.my/Ssouth/SSIntro.htm

Transnational Family Research Institute

Established in 1972, TFRI is a nonprofit research organization in the behavioral sciences with a focus on fertility behavior. Coöperative studies are conducted with colleagues in developed and developing countries.

www.comcast.com

Union of Concerned Scientists

The Union of Concerned Scientists is an independent nonprofit alliance of more than 100,000 concerned citizens and scientists who augment rigorous scientific analysis with innovative thinking and committed citizen advocacy to build a clean er, healthier environment and a safer world.

www.ucsusa.org

United Nations Development Fund for Women

UNIFEM provides financial and technical assistance to innovative programs and strategies that promote women's human rights, political participation and economic security. Within the UN system, UNIFEM promotes gender equality and links women's issues and concerns to national, regional and global agendas by fostering collaboration and providing technical expertise on gender mainstreaming and women's empowerment strategies.

www.unifem.org

Population Studies Center, University of Michigan

The University of Michigan's Population Studies Center, originally established in 1961 as a unit within the department of sociology, has had close connections to the department of economics since 1966. The center has become increasingly interdisciplinary over time, drawing faculty from sociology, economics, anthropology, public health and public policy. The energy and intellectual curiosity of the center's researchers, fostered by the strong support environment and leavened by their interaction with visitors and students at all levels, is a major source of the center's momentum. The PSC is comprised of independent population researchers who pursue their own agendas with the support of the PSC cores. A large portfolio of both domestic and international research is strong in several key areas of demographic research: 1) family formation, fertility, and sexual behavior; 2) inequality, human capital, race and ethnicity; 3) health, disability and aging; 4) migration and population dynamics; and 5) education, training and methodology.

www.psc.isr.umich.edu

Carolina Population Center, University of North Carolina

The Carolina Population Center is a community of scholars and professionals collaborating on interdisciplinary research and methods to advance understanding of population issues. Based in Chapel Hill, the center extends its resources to path-breaking work in the U.S. and 50 other countries, making its findings available to a global audience. A nationally recognized training program educates the next generation of population scholars.

www.cpc.unc.edu

Women's Environmental and Development Organization

WEDO is an international organization advocating for women's equality in global policy. It seeks to empower women as decision-makers to achieve economic, social and gender justice plus a healthy, peaceful planet and human rights for all.

www.wedo.org

World Health Organization

The World Health Organization is the United Nations' specialized agency for health. Established in 1948, WHO's objective, as set out in its constitution, is the attainment by all peoples of the highest possible level of health. Health is defined in its constitution as a state of complete, physical, mental and social wellbeing, not merely the absence of disease or infirmity. WHO is governed by 192 member states through the World Health Assembly composed of representatives from WHO's member states who approve their program and budget for the following biennium and decide major policy questions.

www.who.int

World Leaders Statement in Support of ICPD

The World Leaders Statement in Support of the International Conference on Population and Development (ICPD) succinctly reaffirms the commitment to implement the "Program of Action" agreed to by 179 nations in Cairo, Egypt in 1994. The ICPD action plan is a 20-year vision for health and human rights, women and development. The purpose of the World Leaders Statement is to reaffirm the ICPD's vision at its midpoint, to galvanize support in the public and private sectors and to generate new momentum for achievement of ICPD goals.

www.icpdleadersstatement.net

World Resources Institute

World Resources Institute is an independent nonprofit organization with a staff of over 100 scientists, economists, policy experts, business analysts, statistical analysts, mapmakers and communicators working to protect the earth and improve people's lives.

www.wri.org

World Wildlife Fund

The World Wildlife Fund leads international efforts to protect endangered species and their habitats. Now in its fifth decade, WWF works in more than 100 countries around the globe to conserve the diversity of life on earth. With nearly 1.2 million members in the U.S. and another four million worldwide, WWF is the world's largest privately financed conservation organization. It directs its conservation efforts toward three global goals: saving endangered species, protecting endangered habitats and addressing global threats such as toxic pollution, over-fishing and climate change. From working to save the giant panda and bringing back the Asian rhino, to establishing and helping manage parks and reserves worldwide, WWF has been a conservation leader for more than 40 years.

www.wwf.org

Additional copies of this book may be obtained
from your bookstore
or by contacting
Hope Publishing House
P.O. Box 60008
Pasadena, CA 91116 - U.S.A.
(626) 792-6123 / (800) 326-2671
Fax (626) 792-2121
E-mail: hopepub@sbcglobal.net
www.hope-pub.com